Things that Go BOOM

or Fly, Float, and Zoom!

Things that Go BOOM or Fly, Float, and Zoom!

ALAN AND GILL BRIDGEWATER,
GLYN BRIDGEWATER, JULIAN BRIDGEWATER,
PAUL CLARK, IAN LAMBERT, MIKE RIGNALL

St. Martin's Griffin
New York

THINGS THAT GO BOOM OR FLY, FLOAT, AND ZOOM!

Printed in Thailand. For information, address St. Martin's Press, 175 Fifth Avenue, New York, N.Y. 10010.

www.stmartins.com

Library of Congress Cataloging-in-Publication Data
Available Upon Request

ISBN: 978-0-312-57404-8

First U.S. Edition: November 2009

10 9 8 7 6 5 4 3 2 1

Printed in Thailand

This book was conceived, designed and produced by

IVY PRESS
Ivy Press
210 High Street
Lewes
East Sussex BN7 2NZ
United Kingdom

CREATIVE DIRECTOR: Peter Bridgewater
PUBLISHER: Jason Hook
CONCEIVED BY: Sophie Collins
EDITORIAL DIRECTOR: Tom Kitch
SENIOR PROJECT EDITOR: Polita Caaveiro
ART DIRECTOR: Wayne Blades
DESIGN: Martin Topping
PHOTOGRAPHER: Andrew Perris
ILLUSTRATOR: Richard Palmer

The publisher wishes to thank the models who took part (in order of appearance): Paul Clark (Pyrotechnic Rocket); Andrew Lawes and Joshua Furminger (Lemon Cannon); Harley Bridgewater (Heli-Whirler); Julian Bridgewater and Glyn Bridgewater (Fire Kite); Nina Bennett-White (Paddle-Operated Punt); Joseph Bridgewater and Jessica Bridgewater (Pneumatic Boat); Julian Bridgewater and Glyn Bridgewater (Hot Air Balloon); Nina Bennett-White and Isla Double (Free-Flight Airplane); Harvey Bridgewater, Nina Bennett-White and Jessica Bridgewater (Soapbox Go-Cart); Alan Bridgewater (Drum-Beating Whirligig).

Contents

Introduction

If you like making things—or if you're someone who would like to make things but who never quite gets around to it because you're not absolutely sure what you'd like to make—this book was put together with you in mind. The projects in *Things That Go Boom or Fly, Float, and Zoom* vary from the retro to the avant-garde, with a few excursions into the frankly oddball. They're the combined result of seven inventive minds going to work on a basic brief: To create (or recreate) some ideas purely for fun.

Although none of the machines you'll find here will split the atom or even fetch you a cold beer from the refrigerator, every one of them does something: floats, flies, fires, walks, or "talks". Each of them will keep you occupied for an hour or two (or even a day or two); you can turn some of them into group affairs (in fact, don't even try launching the Hot Air Balloon without a friend), while others are one-person enterprises (you'll want the satisfaction of getting your Electronic Didgeridoo into perfect tune before unveiling it to an admiring audience). If you've always dreamed of impressing the kids, the little balloon boat seems to fascinate toddlers, while the traditional Go-Cart and the Lemon Cannon were unfailing hits with slightly older testers. There really is something for everyone here.

The projects call on a range of skill levels, from very basic gluing and folding to more demanding woodworking, soldering, and circuit-board construction. If you're a complete beginner to a specific technique, why not try out one of the simpler projects first? (If you're new to woodworking, for example, start off with the Pneumatic Boat or the Heli-Whirler instead of the more complex Whirligig or the Paddle-Operated Punt.) If you favor Stomper or the Electronic Musical Box, check out the detailed notes on building a circuit board on pages 8–11 first. They've been written for complete newcomers to electronics and should help you through to a successful conclusion.

Whichever project you choose, read it from beginning to end before you start. This will help ensure that you have enough time to finish it once you've started. It's better to make something simple that you know you'll finish that day (or at least that weekend) than to aim higher and end up with a half-made machine that lurks reproachfully in the corner. However, don't be overcautious—take an excursion at least a little outside your comfort zone. If you've always stuck to woodwork, try something electronic; if you're comfortable with electronics but aren't used to a world of paper and glue, try making the Hot Air Balloon or the Fire Kite. You may find that you can painlessly extend your skills at the same time as you're enjoying yourself.

Finally, don't neglect the safety advice. Making machines, even straightforward ones, is never entirely risk-free and we've highlighted specific safety issues where relevant. In particular, the machines that float and fire need careful attention to detail, so proceed cautiously to ensure the maximum amount of fun both making and using your chosen machines.

ABOVE AND RIGHT One of the more complex projects featured in the book, the Light-Following Robot.

Constructing Circuit Boards

Many of the projects involve making circuits. Here we are going to explain how to construct your own circuits from the diagrams featured in the book.

Things You Need

The electronic parts used in the projects are very common and are easy to buy. You can order them from reputable online dealers or from electronics stores.

CIRCUIT BOARD Two common types of circuit board are known as stripboard and matrix, or perforated, board. Both consist of a grid of holes spaced ⅛ inch (2.54mm) apart; stripboard has parallel copper tracks that link rows of holes running the length of the board on one side while matrix board has an isolated copper area around each hole. Both have advantages and disadvantages. Stripboard's copper tracks will join two points on a circuit without a wire being needed. The disadvantage is that unless you remember to cut the tracks with a track cutter at the correct points, stripboard will automatically join two points that you do not require. Matrix board requires you to insert each wire, but you do not have the problem of accidental connections.

TRACK CUTTER This is a small hand tool that is used to make cuts in the copper track. If you can't get one, then a large sharp drill bit will work well.

SIDE CUTTERS A small pair will be best so that you are able to trim the excess wire close to the solder joints.

WIRE STRIPPERS You will need these to prepare the plastic-coated wires before soldering. Some strippers can strip many different wire sizes and you must adjust a pressure screw accordingly. Some have sets of holes to choose from, depending on your wire size. If you cannot get a pair, then it is possible to strip wires with side cutters as long as you don't cut through the conductors inside the wire.

SOLDER You can buy solder with different melting temperatures, different diameters, and also different mixes of materials. A regular electrical solder will melt about 370°F (188°C), and it is probably a good idea to get one with a diameter of around 0.028 inches (0.7mm). This is commonly referred to as 21 AWG (American Wire Gauge) or 22 SWG (Standard Wire Gauge).

SOLDERING IRON Soldering irons can come in different forms. The cheaper ones plug straight into an electrical outlet and the tip temperature is set by a thermostat inside the tip, or you can have an adjustable setting in the iron. The more elaborate setups have a separate power box where the tip temperature is either set within the tip or is adjustable on the power box. There are also options for the type of tip: some are pointed and others have a flat face at the end. For these projects, set the iron temperature for regular solder and use a tip that is fairly small. If the tip has a flat end, go for one that is about 1⁄16—3⁄32 inch (1.6–2.0mm) across the flat.

DAMP SPONGE OR COTTON PAD These are used for cleaning the iron at regular intervals. Before making each joint, apply a little solder to the tip of the iron and then clean it off on the damp pad. If you do this each time you use the iron (it should have a shiny, silvery tip), you will find that it is easier to heat up the joints and melt the solder.

HACKSAW AND FILE You will probably want to cut the circuit board to a size that suits your project, so it is best to use a small hacksaw and then clean the ends with a file. Beware of loose pieces of leftover copper after the cutting, which can cause short circuits.

WIRE You will need some insulated wire to make the connections between the circuit board and the rest of the parts. For the projects within this book, 20 AWG (0.0320 inch/0.812mm) conductor size wire with PVC jackets will be suitable. Multistrand wires are more flexible.

TINNED COPPER WIRE When you design your board, you may need to link some of the copper tracks together. You can do this with insulated coated wire. A quicker and neater alternative is to use tinned copper wire that is ready to solder. If you buy a roll, choose one that is around 21 AWG (0.028 inches/0.7mm).

Laying Out the Circuit

It's a good idea to start by sketching your circuit layout onto paper. First of all, have a good look at the circuit and see if you can see a structure that is going to work. Remember that if two lines in the circuit cross and there is a black blob at the crossing point, this indicates that there is supposed to be an electrical connection between those lines. Remember that if you have a number of components connected to a line, it doesn't matter where or in which order they connect to each other.

Start by placing any ICs (integrated circuit) so that the rows of pins run at right angles to the direction of the copper strips. If you are using stripboard, you will need to cut the copper tracks in the middle of each IC to separate the sides. Next, add your links and the rest of your components to the sketch and keep them all parallel to each other and facing at right angles to the copper track. When you have an idea that you think works, try laying the parts in their respective positions on the board to see how they all fit together (this should give you an idea of how big the board needs to be). Don't forget to assign holes on the board for connecting the wires. You may also need to allow space to make mounting holes in the board for screwing it in place in your machine's build.

Preparing the Board

Once you have worked out the size of your board, you can then cut it out using a hacksaw. Tidy up the edges with the file and drill any necessary mounting holes. If you are using stripboard, make all the required track breaks using the track cutter and make sure there are no small parts of copper that may come loose and create short circuits between the tracks. You may also need to use the track cutter to break the copper track around the mounting holes to ensure you don't create any short circuits with the securing screw. If you are using matrix board, you will need to insert all the wires, but no track cutting is required.

BELOW An example of a bird's-eye view of a circuit board, showing all the components, which when connected give each electronic machine its brain.

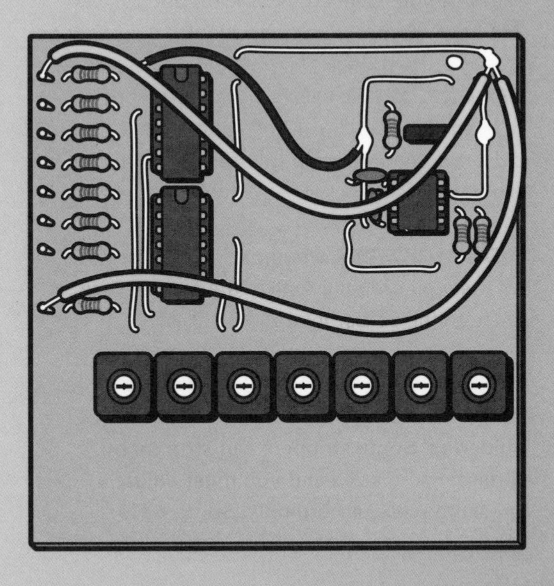

Electronic Component Symbols

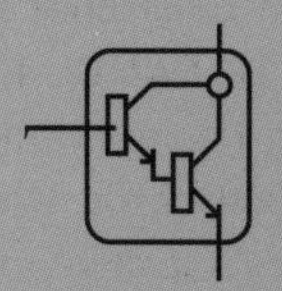
Transistor (Darlington)

Connection

Resistor

Adjustable resistor

Capacitor

Electrolytic capacitor

Terminal pin

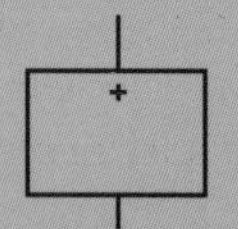
Power source

Motor

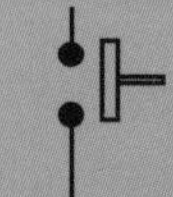
Switch (push-button)

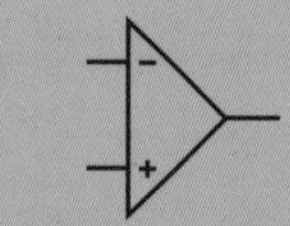
Operational amplifier

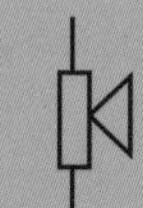
Loudspeaker

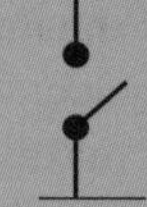
Switch (on/off)

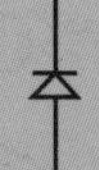
Diode

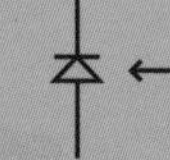
Photodiode

Light emitting diode

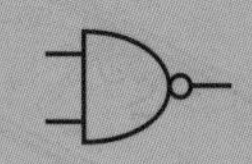
NAND logic gate

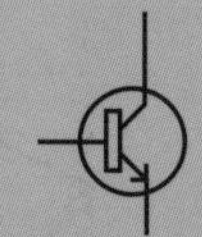
Transistor (NPN)

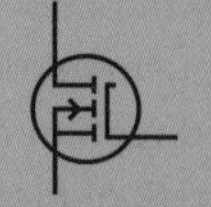
MOSFET

Caution

When soldering, remember that the iron can be as hot as 750°F (400°C), so make sure you store it somewhere safe while it is on. Some components, such as ICs, transistors, and diodes, are more sensitive to temperature than others, making it possible to damage them by overheating them when soldering onto them. Make sure you hold the iron on to them for the minimum amount of time.

Ideally, you should ground yourself, for example, with a water pipe or a conductive wrist strap, to avoid problems with static electricity, which can damage some components.

Soldering the Parts

It's a good idea to start by soldering the ICs in first, because it is easy to hold them in place on the board when there are no other components by simply turning the board upside down (on top of them). Wet the iron tip with solder and clean it using the damp sponge. Position the tip of the iron so that it makes contact with both the board and the component lead, and then immediately start trying to feed the solder in. As soon as the solder flows across the copper, remove the iron. If the iron is not held on for long enough, the solder will sit like a blob on the board and will not make good contact. If the iron is held on for too long, there is a risk of creating a dry joint, where all the bonding component of the solder has time to boil off.

Next, place the other components on the board one-by-one or in groups and pass the legs of the components through the board. You can put a small bend in the leg of the component to make it stay in place while you solder it in and if you are using stripboard, it is best to bend it in the direction of the copper strips to reduce the risk of creating short circuits. Solder the components in place and then snip the ends close to the solder joints.

Connecting the Wires

Before you solder any wires onto the board, they must be prepared. Strip around ½-inch (12mm) of the plastic insulation off the wire, then twist up the conductors if you have a multistrand wire. Touch the soldering iron onto the bare wire and add solder until it flows all over the conductor (this is called tinning). Now pass the wire through the board from the noncopper side and solder it in the same way as previously for the other components. Be aware that the jacket on some wires can shrink a little when it gets too hot, so be careful not to overheat it.

Desoldering

At some stage you may need to desolder a joint to remove or reposition a wire or component. Use either a desoldering pump or a solder remover wick, which are both available at hardware stores.

Final Inspection

Check the copper side of the board very carefully for any pieces of loose solder or shorts between tracks. You may see blobs of brown material around the solder joints. This is the bonding flux used in the solder and it is fine to leave it.

BELOW When soldering, be careful to avoid touching the electrical cord with the tip of the iron.

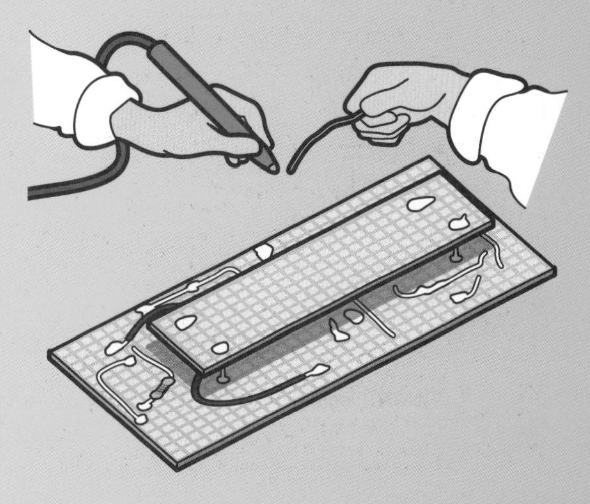

Safety

Every project within this book has been created with a view to making it as safe as possible for the reader, but whatever you try from this book you try at your own risk. The authors, the publisher, and the bookseller cannot and will not guarantee your safety. Obeying the following safety points will help to reduce, but will not eliminate the chances of accidents occurring.

- The projects are intended to be made and used by adults. Children should not attempt to make them, and they should only use them under close adult supervision.
- Always read the instructions and safety warnings carefully before you start making or using a project.
- Always read the usage instructions of any tool or component you intend to use. Store tools in a safe place, out of the reach of children and pets.
- Wear safety goggles at all times when making and testing the projects. Many projects will also instruct you to wear additional safety protection, such as gloves or a dust mask.
- Do not substitute materials or tools, or make alterations to any of the construction techniques described in the step text.
- When using a power drill, always unplug the drill before attaching new parts and remove the chuck key before switching it on. Do not wear loose clothing or jewelry that could get caught.
- Check and follow all local, state, and federal regulations.
- Carefully plan the project before you begin work on it. Be aware of your limitations, and if in doubt, ask an experienced person for help.
- Use your common sense—if you have any doubts at all, then stop work immediately.

MACHINES THAT TURN, THROW, AND FIRE

Six-Volt Wind Turbine

The wind turbine was inspired by vintage photographs taken of do-it-yourself turbines built in France during World War II, used as mini-generators to power everything from a lightbulb to a butter churn. Our hub is a recycled dyno hub from a bicycle, and the propeller blades are made from a thin aluminum sheet.

Mounted on a tall pole (or rooftop), this domestic windmill will supply you with enough power to light up your workshop while you're working on other machines. You can adapt it to power something else, if you prefer. There's no reason it won't charge your cell phone, or even work as backup to outdoor solar lights.

YOU WILL NEED

- 30-inch- (760-mm) length of 4-inch (100-mm) diameter plastic drainage pipe (to join paddles to paddle wheel blocks)
- Piece of best-quality multicore plywood, 24 × 12 × ¾ inches (600 x 300 x 20mm)
- Twelve common or finishing nails
- 6-volt dyno hub with three nuts to fit the spindle (we used a dyno from a bicycle, but you can also use a small generator from a washing machine or moped)
- Tissue
- Sixteen 2-inch (50-mm) bolts, ⅜ inch (10mm) in diameter, with nuts and washers to fit
- Piece of aluminum sheet, 24 × 16 × 1/16 inches (600 × 400 × 2mm)—this allows for cutting waste
- Two pieces of prepared knot-free pine, 19 × 2 × ¾ inches (480 × 50 × 20mm)
- Steel bracket to fit your chosen dyno hub (we got ours from a farm supplier—part of a wheel unit designed to fit a large-swing farm gate)
- Eight 1½-inch (40-mm) screws
- Four guy ropes
- Four pegs to attach guy ropes
- Two hose clamps
- Length of ¾-inch- (20-mm) diameter metal tube (the type of tube used for backyard awnings)—the height, metal gauge, and setup will need to relate to your location
- Heavy duty washer
- 2-core lightweight cable (the type used for bicycle and battery electrics), its length to suit the height of your mast, and the distance of the mast from your shed, caravan, or whatever
- One 6-volt bicycle light

TOOLS

- Compass (geometrical)
- Cordless drill/driver with a good selection of drill bits to fit
- Electrical insulating tape
- Hammer
- Heavy-duty gloves
- Large pair of tin snips
- Masking tape
- Medium-size file
- Pencil, ruler, and square (for measuring and drawing lines)
- Pliers
- Saber saw with a good selection of "medium" blades to fit
- Sandpaper (fine grade) and sanding block
- Straight handsaw
- Workbench with vise
- Wrenches to fit your chosen nuts and bolts

How to Make the Six-Volt Wind Turbine

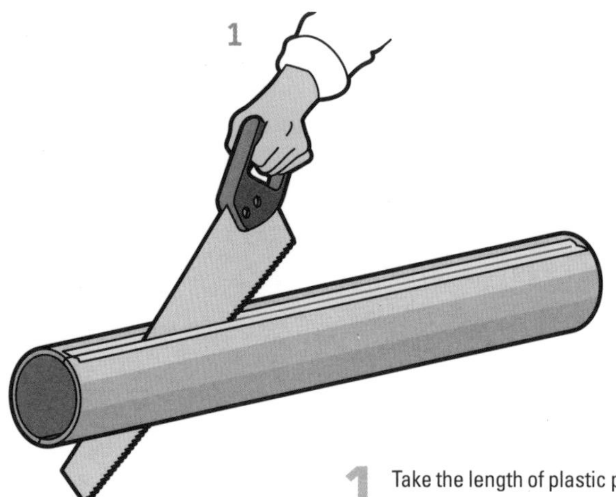

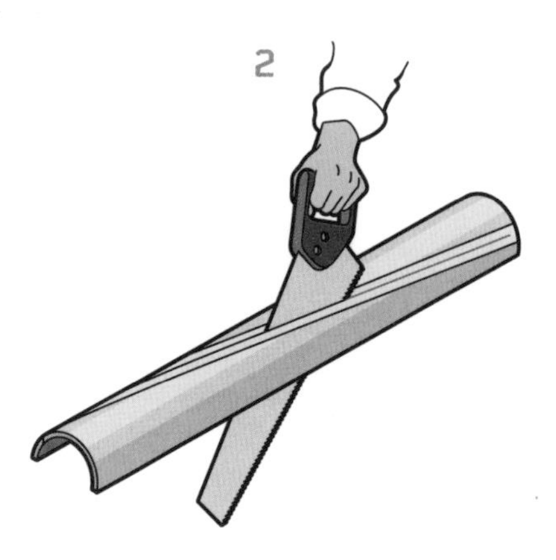

1 Take the length of plastic pipe and place a strip of masking tape along the center of its horizontal axis. Pencil a line down the center of the tape. Using the straight handsaw, cut the pipe in half along the pencil line, so that you have two identical half-cylindrical shapes.

2 Starting 3 inches (75mm) from one of the corners, run a strip of masking tape along the outside of one piece of the cut pipe to the opposite corner. Repeat this with the other piece. Pencil a line, as previously, down the center of the tape. Again using the straight saw, cut down the lines to create four identical 30-inch- (750-mm) long propeller-blade shapes.

3

3 Pencil a line down the center of the plywood and saw through so that you have two 12-inch (300-mm) squares. Pencil cross diagonals on one of the plywood squares, then position a compass at the intersection, set it to a radius of 5 inches (125mm), and inscribe a 10-inch- (250-mm) diameter circle.

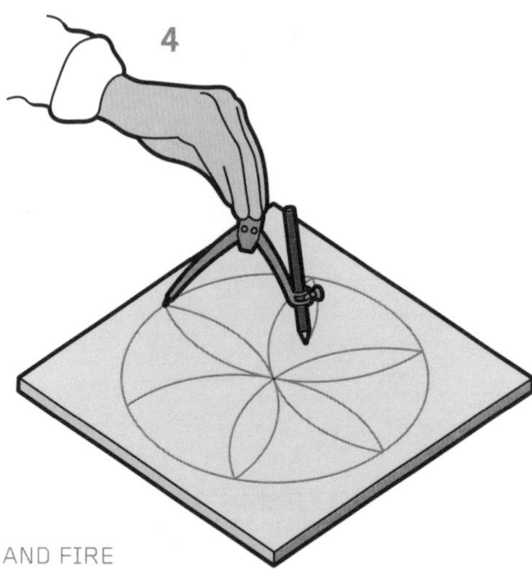

4 Keeping the same radius throughout, reposition the point of the compass anywhere on the circumference of the circle and scribe an arc that intersects both sides of the circle, then position the compass at one of the intersections and scribe another arc. Repeat this until you have drawn out a six-petal shape.

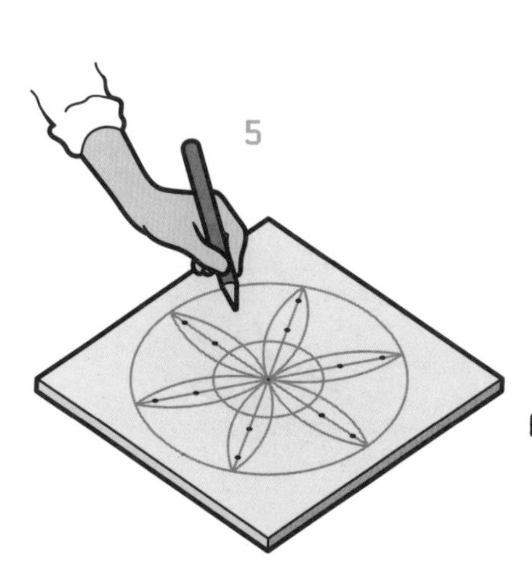

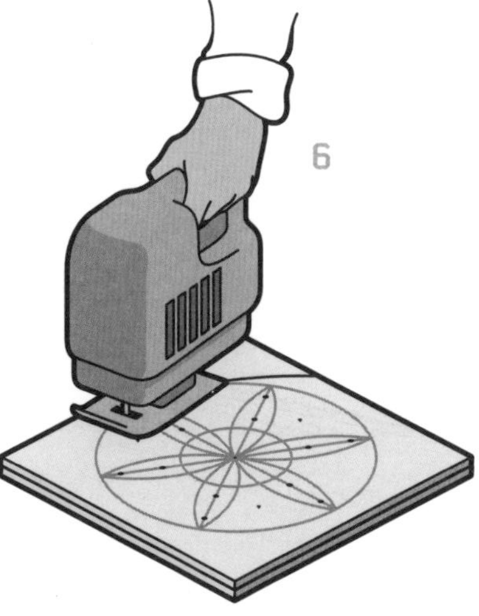

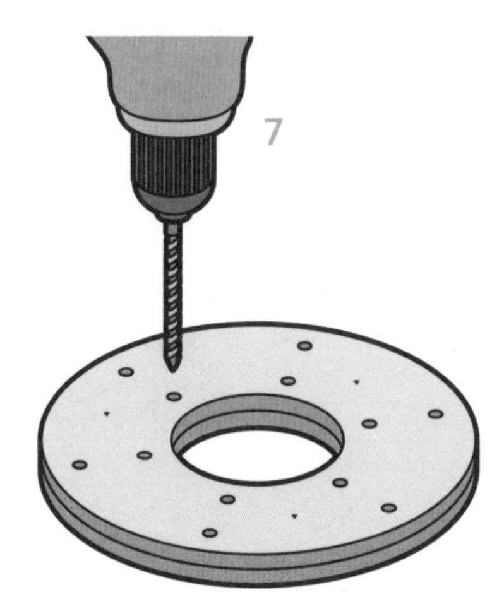

5 Reset the compass to a radius of 2 inches (50mm), spike the point on the center of the petals and draw a 4-inch- (100-mm) diameter circle. Draw straight lines from the center of the circle through to the petal points—so that each petal is halved along its axis—and along the line drawn through each petal, mark two points, one 2¾ inches (70mm) from the center of the circle, the other 4¼ inches (110mm). You'll end up with 12 points in all (these mark where the bolt holes will be made).

6 Put two sheets of plywood one on top of each other—so that the marked-out piece is uppermost—and secure them together with finishing nails, hammering these in but leaving the heads protruding so that you can remove them at a later stage. Using the saber saw, cut around the circle so that you end up with a disk.

7 Drill a pilot hole through the center of the disk (large enough for the saber-saw blade to pass through), unhitch the blade of the saber saw, position it inside the hole, rehitch the blade, switch on, and cut around the inner circle. You should finish up with two identical 10-inch-(250-mm) diameter disks—still nailed together—each with a 4-inch- (100-mm) diameter hole at the center. Use the drill and bit to bore out the eight bolt holes.

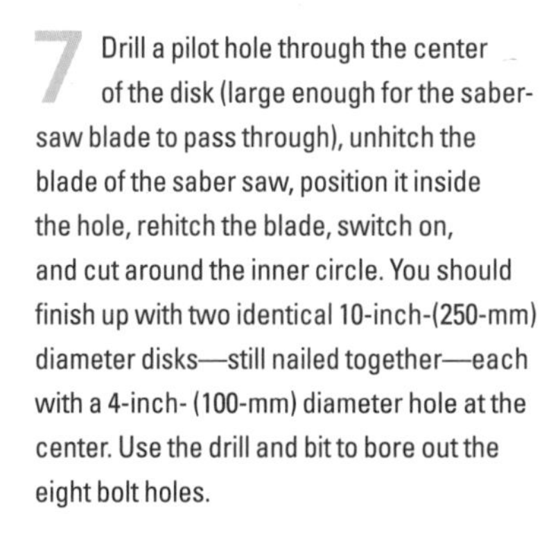

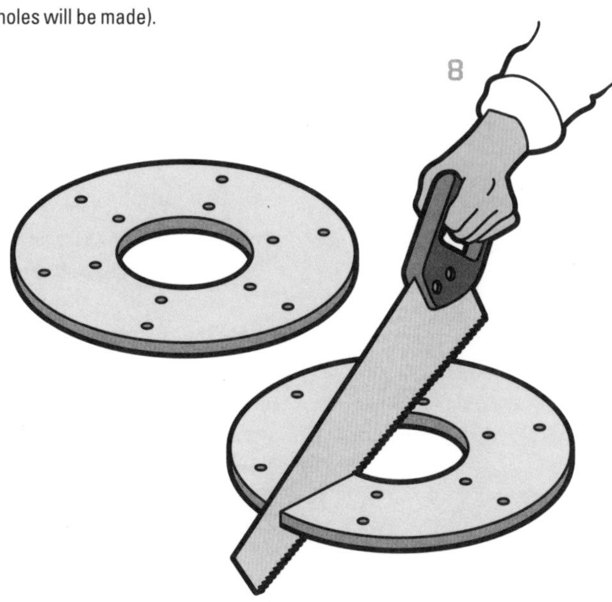

8 Ease the disks apart, remove the finishing nails, and then use a straight handsaw to run a saw cut from the edge of one of the disks through to the central hole. This let will the wood "give" a little when the dyno hub is sandwiched in between.

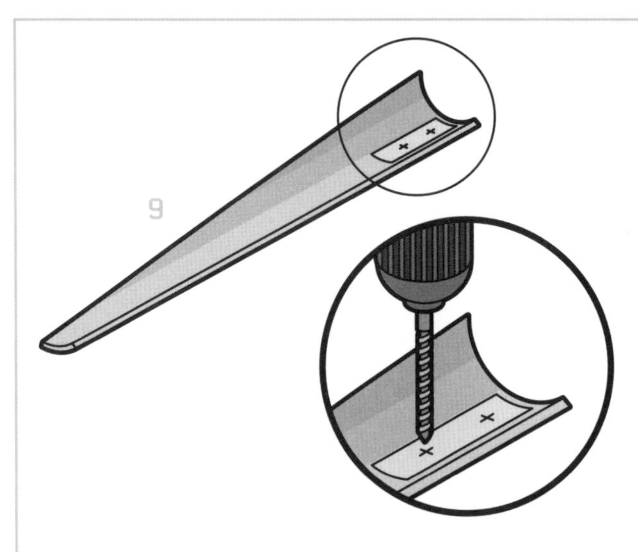

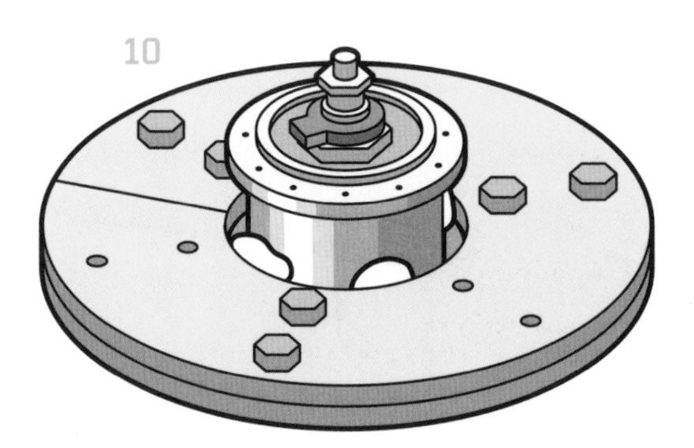

9 Take three of the four propeller blades you made earlier (keep one as a spare), and saw off exactly 6 inches (150mm) from the narrow end of each, and keep the 6-inch (150-mm) narrow-end pieces to use later on. Mark two points, 1½ inches (40mm) apart toward the wide end of the propeller blades (see illustration) and drill out the bolt holes. Ensure that the position of the holes is exactly the same for each blade.

10 Take the dyno hub, making sure that the spindle runs without interference, and then carefully sandwich and pinch the hub rim between the plywood disks, so that the electrical connections on the hub are nearest the rim on the other side of the hub. Wedge little balls of tissue between the hub and the plywood so that the hub is centered and contained, then insert bolts through six of the 12 bolt holes as shown, slide on the washers, screw on the nuts, and tighten the bolts.

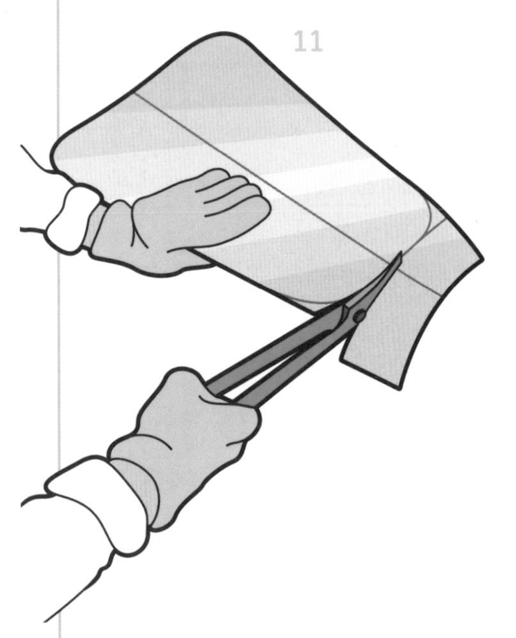

11 Draw out the shape of the tail vane on the aluminum sheet. The shape and measurements of this are up to you, but our one is 21 inches (530mm) long, 14 inches (350mm) wide at one end, and 8½ inches (220mm) at the other. Wear heavy-duty gloves to cut out the shape with tin snips. Use the file to "stroke" the cut edges to a smooth-to-the-touch finish.

12 Take the two 19-inch (480-mm) lengths of wood, and clamp them together in the vise so that the 2-inch- (50-mm) wide face is uppermost. Pencil a centerline running 9½ inches (240mm) along the length, and saw along this to create the two slotted tail vane struts.

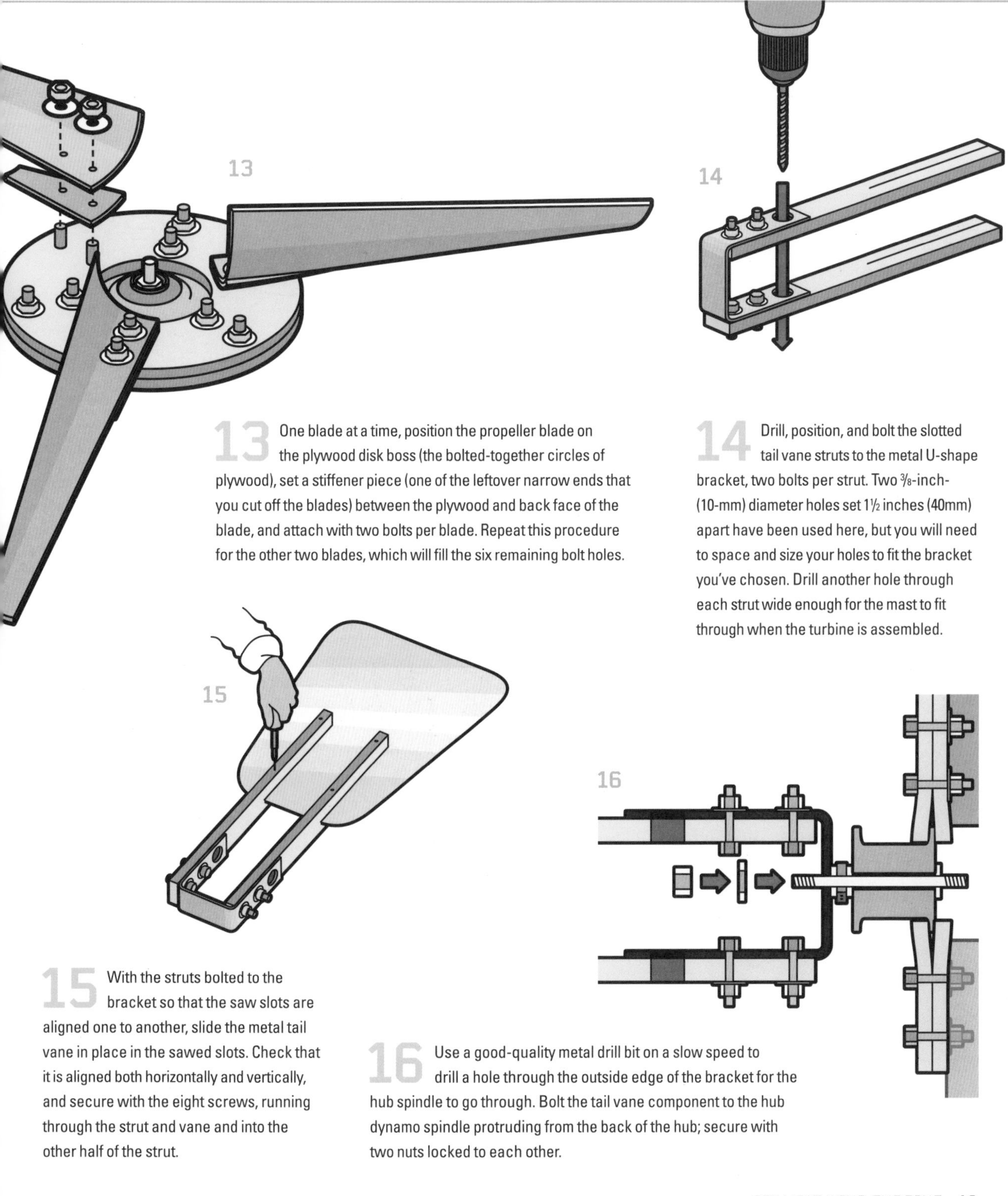

13 One blade at a time, position the propeller blade on the plywood disk boss (the bolted-together circles of plywood), set a stiffener piece (one of the leftover narrow ends that you cut off the blades) between the plywood and back face of the blade, and attach with two bolts per blade. Repeat this procedure for the other two blades, which will fill the six remaining bolt holes.

14 Drill, position, and bolt the slotted tail vane struts to the metal U-shape bracket, two bolts per strut. Two ⅜-inch- (10-mm) diameter holes set 1½ inches (40mm) apart have been used here, but you will need to space and size your holes to fit the bracket you've chosen. Drill another hole through each strut wide enough for the mast to fit through when the turbine is assembled.

15 With the struts bolted to the bracket so that the saw slots are aligned one to another, slide the metal tail vane in place in the sawed slots. Check that it is aligned both horizontally and vertically, and secure with the eight screws, running through the strut and vane and into the other half of the strut.

16 Use a good-quality metal drill bit on a slow speed to drill a hole through the outside edge of the bracket for the hub spindle to go through. Bolt the tail vane component to the hub dynamo spindle protruding from the back of the hub; secure with two nuts locked to each other.

The Wind Turbine

Our wind turbine is a rotating machine that converts kinetic energy from the wind into mechanical energy. The tail vane swings around to force the propellers to point directly into the face of the wind, causing them to rotate. The drive shaft that runs through the dynamo turns on its axis, letting the dynamo convert the turning motion into electricity, which, in this case, is used to power a lightbulb.

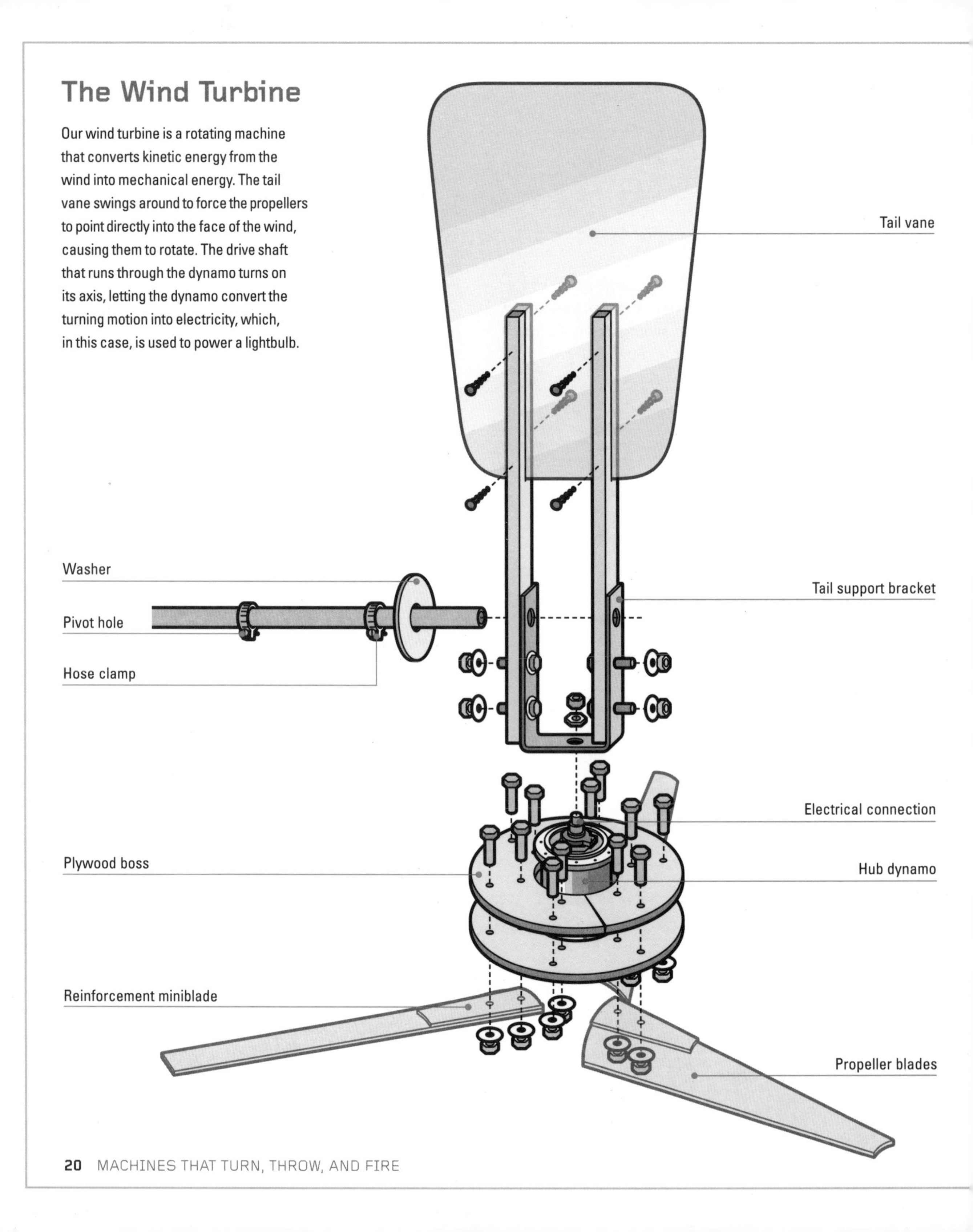

How to Use It

You're almost ready to set up the turbine; all you need to do now is position the mast, slide the turbine in place, choose a device to power, and wait for some windy weather.

Setting Up the Mast

Tie four guy ropes around the pole at the same point, about 3–4 feet (900mm–1,200mm) down from the top, and attach a hose clamp to prevent them from sliding down the pole. With four helpers, each holding a guy rope, and you supporting the base, gently hoist the mast into an upright position, spike it into the ground, and secure each guy rope with a peg. Position the mast at the most suitable point, bearing in mind the height of surrounding buildings and trees, the direction of prevailing winds, and the likelihood of people tripping over the guy ropes.

Attaching the Turbine

Slide the turbine into place on top of the mast, holding it in place with a hose clamp and a heavy-duty washer. Use the pliers and insulating tape to link the two-core cable to the contact points on the turbine. Once this is done, run the cord down the outside of the mast, ensuring you allow a little slack. Finally, link the ground end of the cable to the light (or whatever you've chosen to power with the turbine).

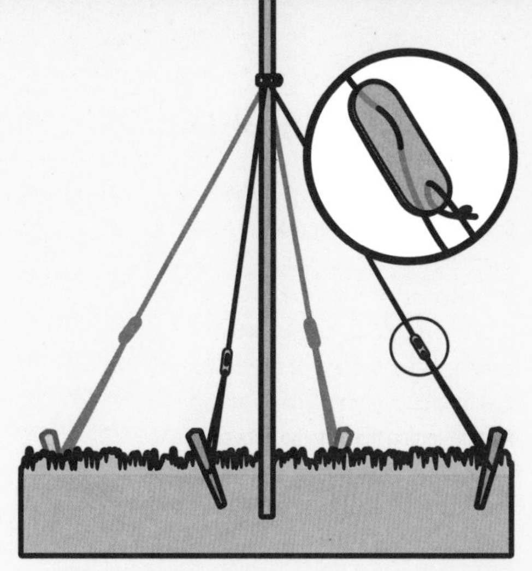

Use the adjustment toggle to make sure the ropes are taut and the pole is vertical.

Slot the pivot pole into the top of the mast.

The turbine is ready to collect energy.

⚠ Safety

- The propeller must be mounted at least 10 feet (3,000mm) above ground level, well away from overhead power lines.
- When setting up the mast, clear a working area 32 feet (10m) around the mast to let it fall over safely.
- The precise details of the wiring will depend on your chosen setup. There is no chance of an electric shock, as the turbine only generates six volts.

Pyrotechnic Rocket

Despite its polished 1950s appearance, the body of our rocket is made from humble materials: it's based on two plastic bottles with a central cardboard tube. Finished in a bright red color with flashes made from holographic tape, balsa-wood fins, and a nose cone and parachute, it fulfills every childhood dream of what a working rocket should look like. Ignition is kept simple: it's made by means of a motor and igniter from a model-making store, so most of the effort in this project goes into the outer construction and appearance.

The rocket flies well and can go impressively high. The parachute ensures a safe return to earth after each flight; without it, the force of the rocket's fall would probably damage its nose or fins, but by adding it, you have the enjoyment of watching your creation float gracefully back down unharmed.

YOU WILL NEED

For the rocket:

- 2-liter and 1-liter plastic beverage bottles (cylindrical, not contoured)
- 24-inch- (600-mm) length of 1⅜-inch- (33-mm) diameter cardboard tube (this is a BT56-size body tube supplied by hobby shops selling model rocket parts, but it can be found at the center of a roll of wrapping paper)
- Kite string
- Thin cardboard sheet (cereal-packaging type, or slightly heavier)
- Small piece of thick cardboard
- Piece of balsa wood, approximately 6 × 40 x ⅛ inches (150 × 1,000 x 3mm)
- Four sheets of thick white paper, 8½ × 11 inches (220 × 280mm)
- Drinking straw
- Heavy kraft paper
- Plastic wrap
- Motor mounting clips, supplied with the motor used in this project (or the flat stainless steel strip from a discarded windshield wiper)
- Deodorant bottle top
- Drinkable yogurt bottle
- Spray paint cap
- 30 ounces (80g) ballast (nails, lead fishing weights, or modeling clay)
- Primer spray paint (car paint)
- Top-coat spray paint (car paint)
- Holographic gift tape

For the parachute and recovery system:

- Black plastic garbage bag
- Eight self-adhesive eyelets
- Kite string
- 12 inches (300mm) of strong elastic
- Aluminum foil

For the launchpad:

- One block of wood or a small plastic flowerpot with cement (or sand and plaster) and grease
- 40-inch (1,000-mm) length of 1/16-inch- (2-mm) diameter steel rod
- Pair of small alligator clips
- 33 feet (10m) of twin core wire; 16½ feet (5m) minimum
- 6v or12v battery
- Doorbell
- Cork to cap the rod for safety

Alternatively, buy an Estes basic launch pad kit from a hobby store, or by mail order.

To fly the model:

- Estes D-12-3 rocket motors and igniters (available from hobby suppliers; the two come together). Model rocket motors (sometimes called "engines") can be sent through the post.

TOOLS

- Adhesive tape
- Cordless drill/driver with a good selection of drill bits to fit
- Cutting board
- Craft knife
- Grease
- Lighter fluid
- Marker pen
- Masking tape
- Metal ruler
- Pencil
- Pliers
- Sandpaper
- Scissors
- Soldering iron (with cored solder and flux)
- Two-tube pack of epoxy adhesive; one tube is the resin, the other is the hardener
- White glue

How to Make the Pyrotechnic Rocket

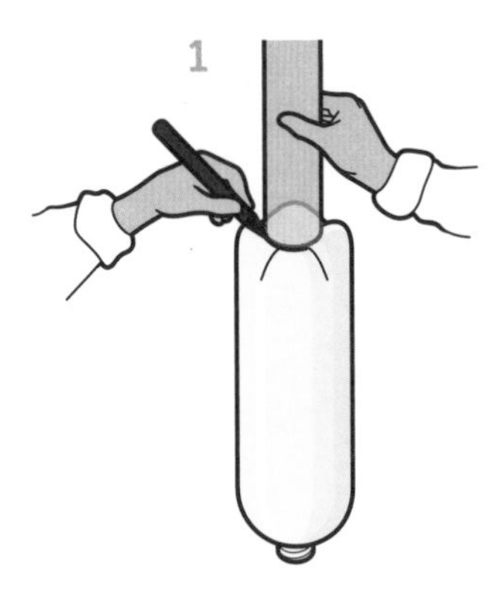

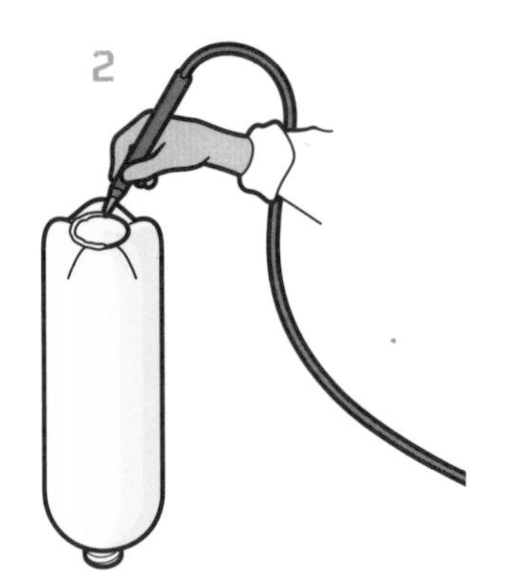

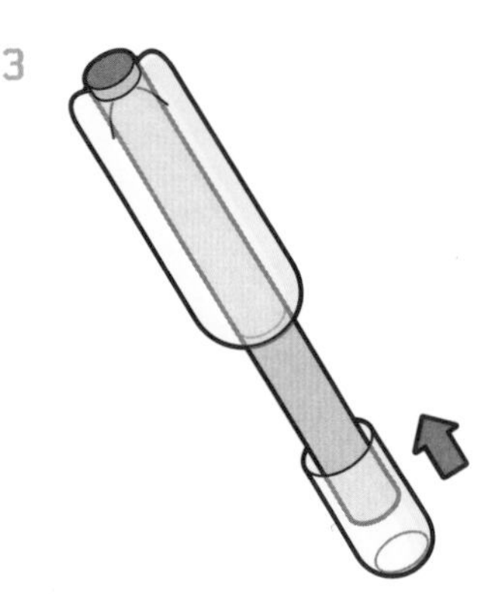

1 Take any labels off the bottles (if they are sticky, use lighter fluid to remove them). Position the cardboard tube in the center of the bottom of the 2-liter bottle, and draw around this with the marker pen.

2 Using an electrical soldering iron, cut a hole by melting the plastic along the drawn line. Alternatively, you can cut the hole out with a knife or a fine saw, but the plastic will be tough so if you try this, be careful! Do the same with the top of the 1-liter bottle and cut the curved bottom end off with a knife.

3 Slide the two bottles onto the body tube, leaving ⅝ inch (15mm) of the tube protruding from one end (the base) and the rest from the other.

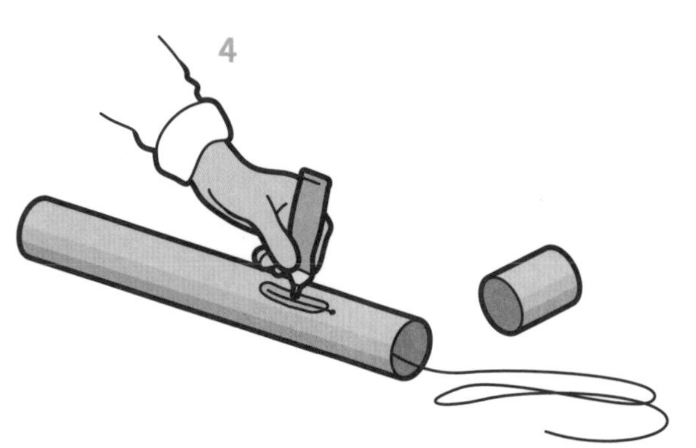

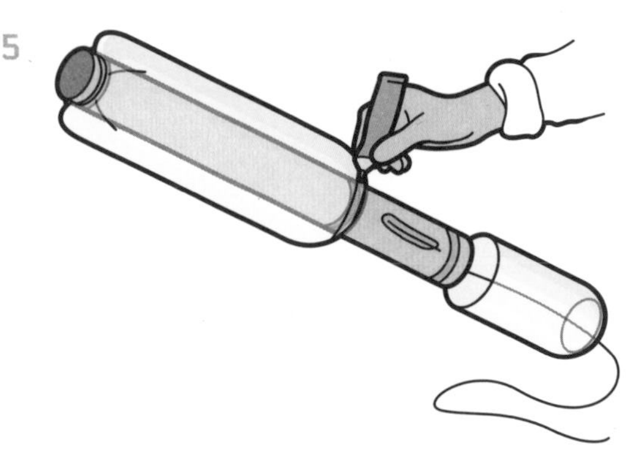

4 Trim the top of the body tube so that ¾ inch (20mm) protrudes above the upper bottle. Remove the bottles from the tube, then make a small hole about 1¼ inches (30mm) from the top of the tube and thread approximately 12 inches (300mm) of kite string through this, leaving 2 inches (50mm) emerging from the outside of the tube (this string will later be attached to the parachute by a length of elastic). Glue this in place with a patch of epoxy adhesive.

5 Now use epoxy adhesive to glue the base of the body tube to the bottom and top of the large bottle, leaving ⅝ inch (15mm) protruding at the base. When the epoxy is dry, slide the top bottle on and glue it to the shoulder of the bottom bottle and the body tube at the top, again with epoxy adhesive. Wipe any surplus away from the top before it sets.

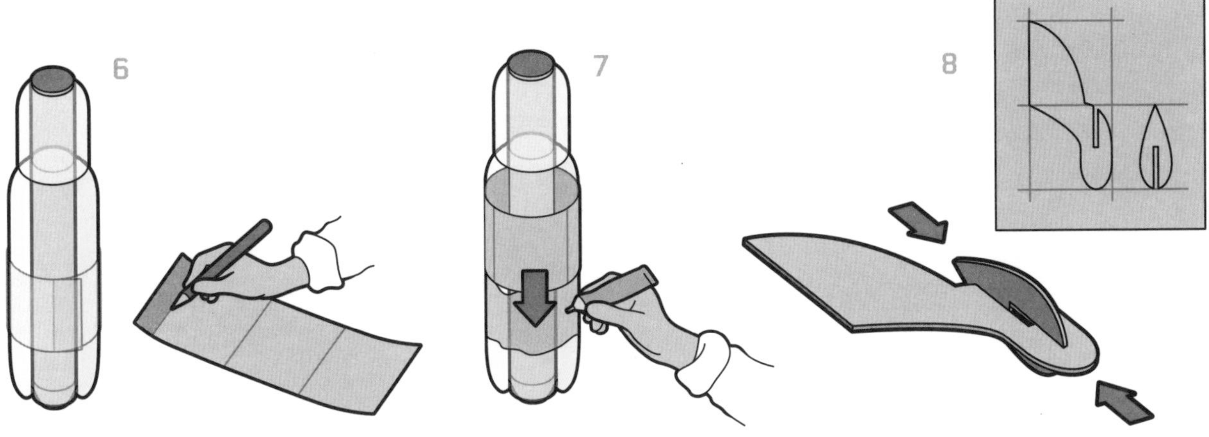

6 Cut a strip of thin cardboard about 14 inches (350mm) long and 4 inches (100mm) deep. Wrap this around the lower part of the 2-liter bottle, just above where the curve of the base starts. Mark the overlap, which should be about 5/8 inch (15mm). Remove the cardboard, lay it flat, measure one-third and two-thirds along the width (excluding the overlap), and mark into sections (each will be a third of the circumference of the bottle).

7 Wrap the cardboard around the bottle, with the lines you've drawn on the outside, and glue the overlap with white glue (tape in place while this is drying). When dry the sleeve should slide up the bottle. Spread some epoxy adhesive thinly on the bottle and slide the sleeve back. Allow to set.

8 Design the shape for your fins, and pencil this onto a piece of stiff cardboard to serve as a template (to use our shapes, see the templates provided on page 152). The fins must be 4 inches (100mm) long in the root and the same swept back from the base. From the tip to the root they should also be about 4 inches (100mm). The design can be curved or straight, as you like. In the design illustrated, a curved design has been used with small additional tip fins at right angles. These are not essential but add to the style of the model. The smaller lozenge-shaped fin is slotted, as is the main one. One slots onto the other.

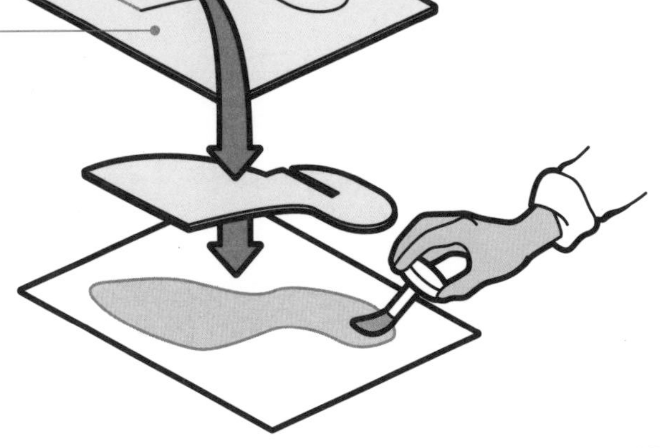

9 Place the cardboard template on the balsa wood and draw around it three times. Cut out the shapes using the craft knife. To strengthen the fins, spread white glue thinly on one side and place the fin onto a sheet of white paper. When dry, turn over and repeat. When dry, trim off the paper around the balsa edge, again using the craft knife.

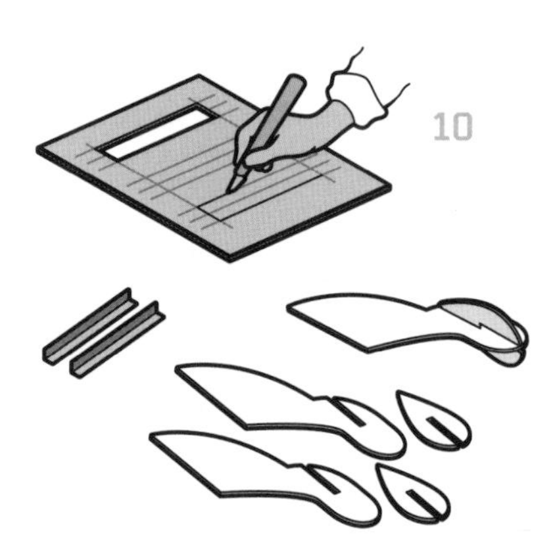

10 On a piece of stiff cardboard, draw three pairs of strips, 4 inches (100mm) long, and 3/8 inch (10mm) wide. Cut these out and fold them down the middle. These will support the sides of the fins and strengthen the joint. The illustration here shows one pair of strips and the three fins. One fin has the tip fin added, the other two haven't been slotted together.

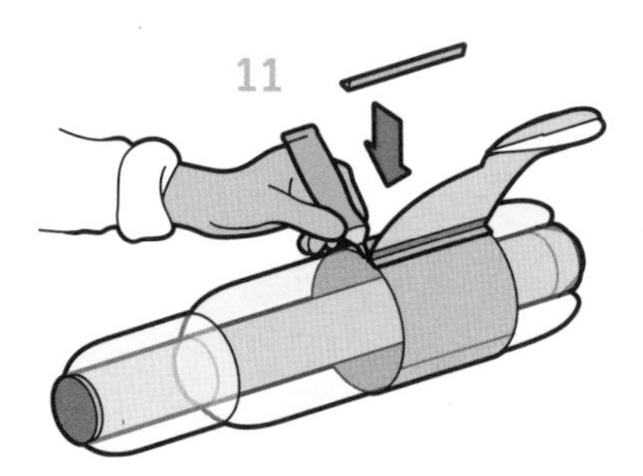

11 Glue the root of one of the fins (using epoxy adhesive) to one of the lines drawn on the sleeve. Add the side strips, also with epoxy. Repeat with the other two fins, keeping the model in place so that each fin is vertical while the adhesive sets.

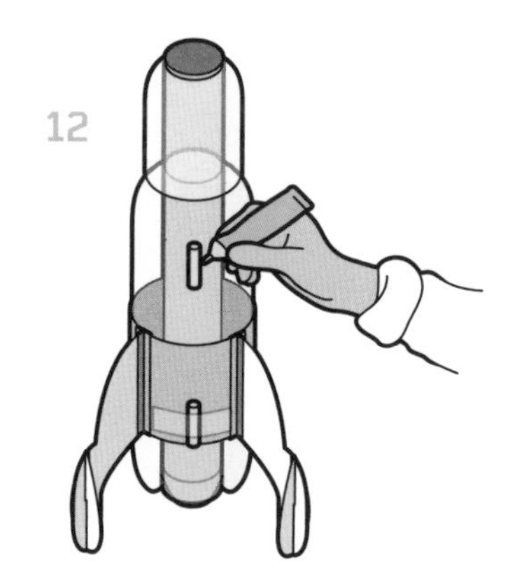

12 To make the launch lugs, cut two pieces off the drinking straw, about 1 1/4 inches (30mm) long, and epoxy these to the larger bottle below the shoulder and at the base of the fin sleeve. When gluing, tape them in place with adhesive tape and leave this on for extra security. The lugs must be in line vertically and centered between two of the fins.

13 To make the motor mount, cut a strip of kraft paper 1 inch (25mm) wide and about 12 inches (300mm) long, using a craft knife and straight edge. Wind plastic wrap around the motor. Spread white glue thinly on the kraft paper and wrap this around the motor. When it is dry, pull the motor out.

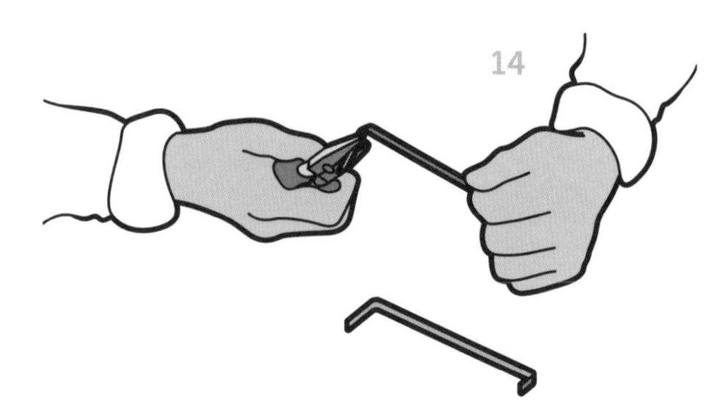

14 If you're unable to purchase a motor mounting kit, make a pair of motor clips from a length of stainless steel strip from a discarded windshield wiper. You will need a strip 3 inches (80mm) long. It can be snapped at the appropriate point by bending backward and forward a few times with pliers. Bend to the dimensions shown, 1/8 inch (3mm) at one end and 1/4 inch (6mm) at the other. The gap between the two bends should be slightly larger than the length of the motor: 2 3/4 inches (70mm). Repeat with the second clip.

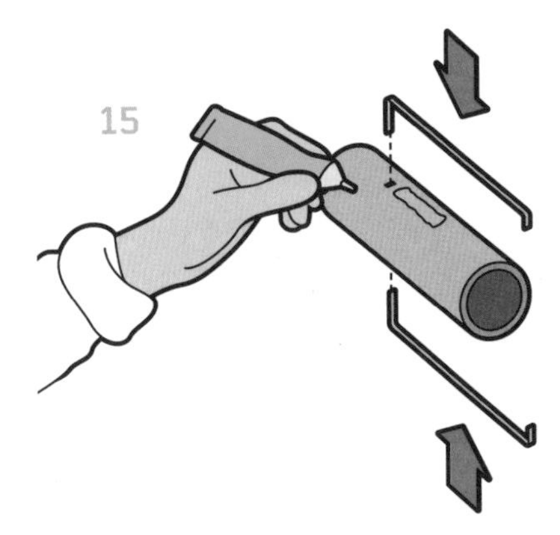

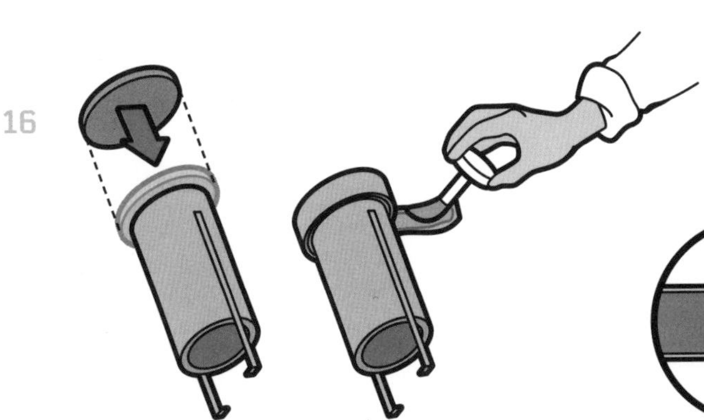

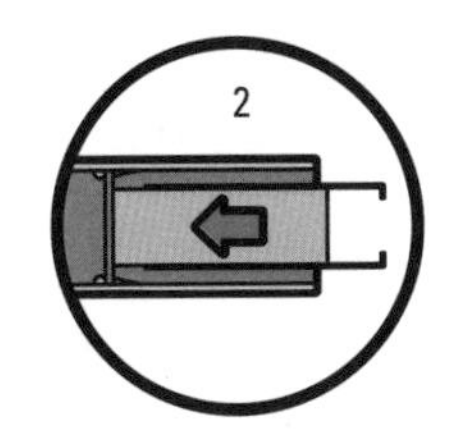

15 Make a small cut ⅜ inch (10mm) from one end of the motor mount. Insert the longer end of the first clip into the cut. Put the second clip on the other side of the mount tube and epoxy both the clips to the mount. Do not glue more than halfway down the tube or the ends of the clip will not be able to flex enough to allow the motor to be taken in and out.

16 To secure the motor mount in the body tube, cut a ring in stiff cardboard just big enough to fit around the top of the motor mount (this is called a centering ring). It's not possible to give precise dimensions because the body tube's internal diameter is variable, depending on what size tube is used. Glue the ring to the top end of the motor mount, and when set, push the mount into the body tube so that the epoxy is pushed up in front of the centering ring like a piston. An alternative to cutting the centering ring, which is tricky, is to wind a band of white-glued paper around the mount tube until it is the correct diameter.

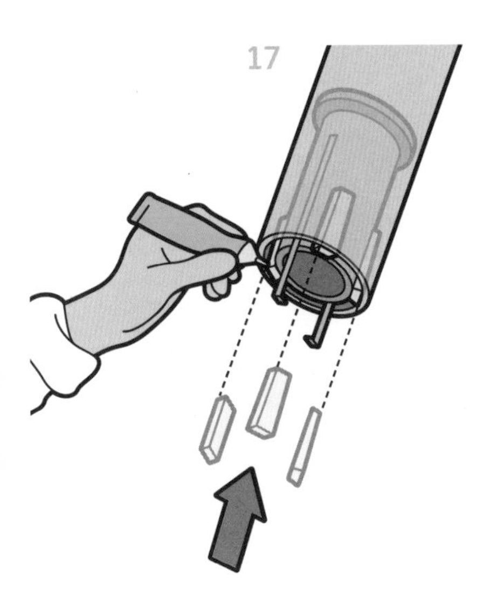

17 Cut three wedges of balsa wood to hold the lower end of the mount in place—again secured with epoxy adhesive. (The exact dimensions will depend on the diameter of the body tube you are using, so some trial and error may be required.)

18 To make the nose cone, you need three components: A conical top nose, a canister that will hold the nose weight, and a lower section that will fit into the top of the body tube and conical top. The canister section needs to be fatter than the body tube. You should be able to find these among everyday packaging. Shown here are a deodorant bottle top, a drinkable yogurt bottle, and a small spray paint cap. The bottle, or cap, needs to fit neatly in the body tube with the clip attached. If you can't find a suitable bottle or cap that fits inside the body tube, take a 3-inch (80-mm) length of body tube and make a connecting piece by cutting a diagonal slice out of it. Fit this inside the body tube. Alternatively, you could use a sleeve of paper.

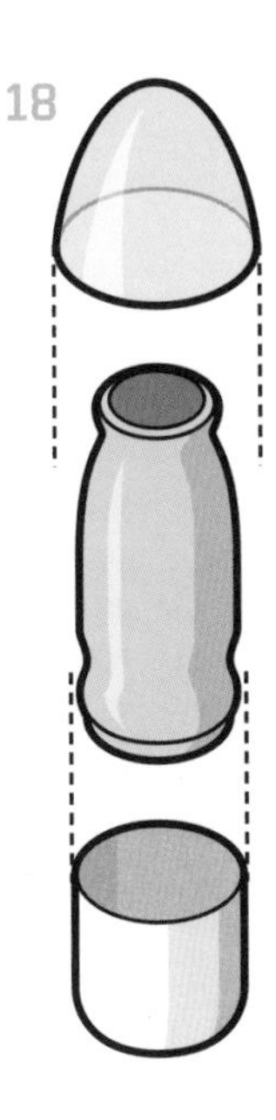

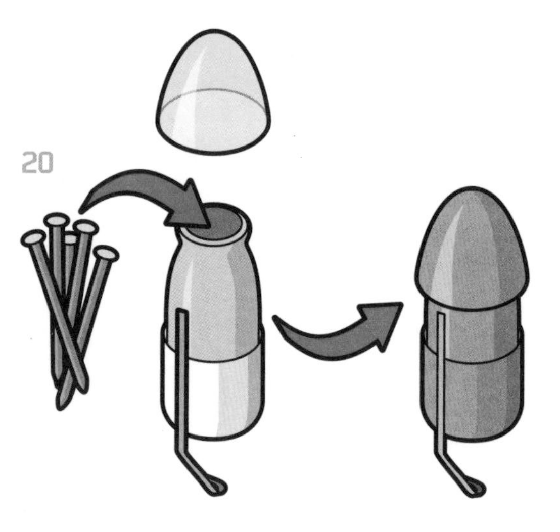

19 Using epoxy resin, glue a clip (made from a 2½-inch- [65-mm] length of the same steel strip used for the motor clip and bent as shown) to the insert that you've chosen as a nose cone (if you do not have a suitable nose cone, you can make one from a cone of cardboard, coated in white glue or epoxy for additional strength).

20 Weight the nose cone (for stability) by gluing in (with epoxy adhesive) nails, lead fishing weights, or modeling clay weighing about 3 ounces (80g). You may also want to paint it, as shown here using gold spray paint.

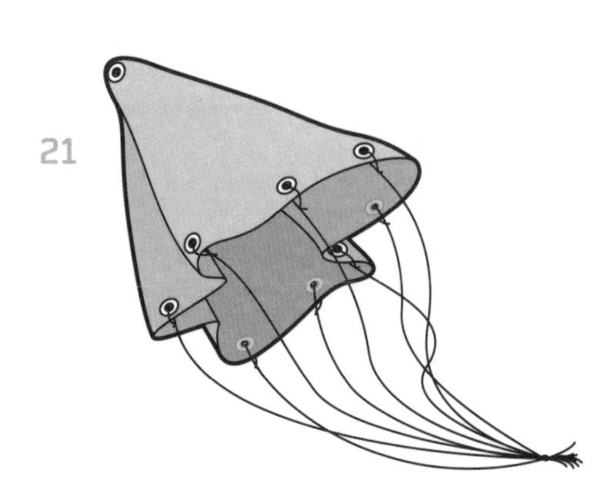

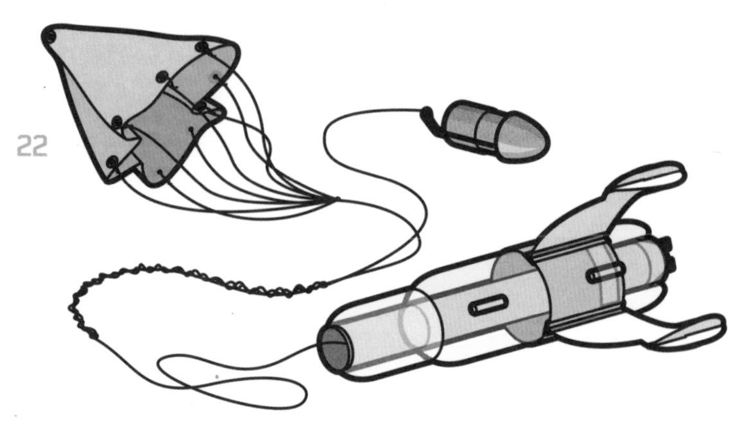

21 To make the parachute, slit the side of the black garbage bag and cut the largest square you can from this. Cut a circle out and attach eight self-adhesive eyelets around the circumference near the edge. Put another in the center. Make a hole in each with a hot wire or by carefully cutting. Tie about 2 feet (600mm) of kite string to each eyelet and tie together at the ends. Fold the parachute into eighths.

22 To make the shock cord, tie one end of the elastic to the nose cone (attaching this to where the eight pieces of string are joined) and the other to the cord extending from the body tube. All the knots must be very tight and double back the elastic so it cannot slip out. A blob of epoxy on any knots will make sure they will not come undone.

23 Following the instructions on the spray paint cans, spray paint the rocket (ensure you are in a well-ventilated space), primer first, if needed, then top coat. When the paint is dry, add further decoration of your choice, using decals (transfers) from other kits and holographic gift tape.

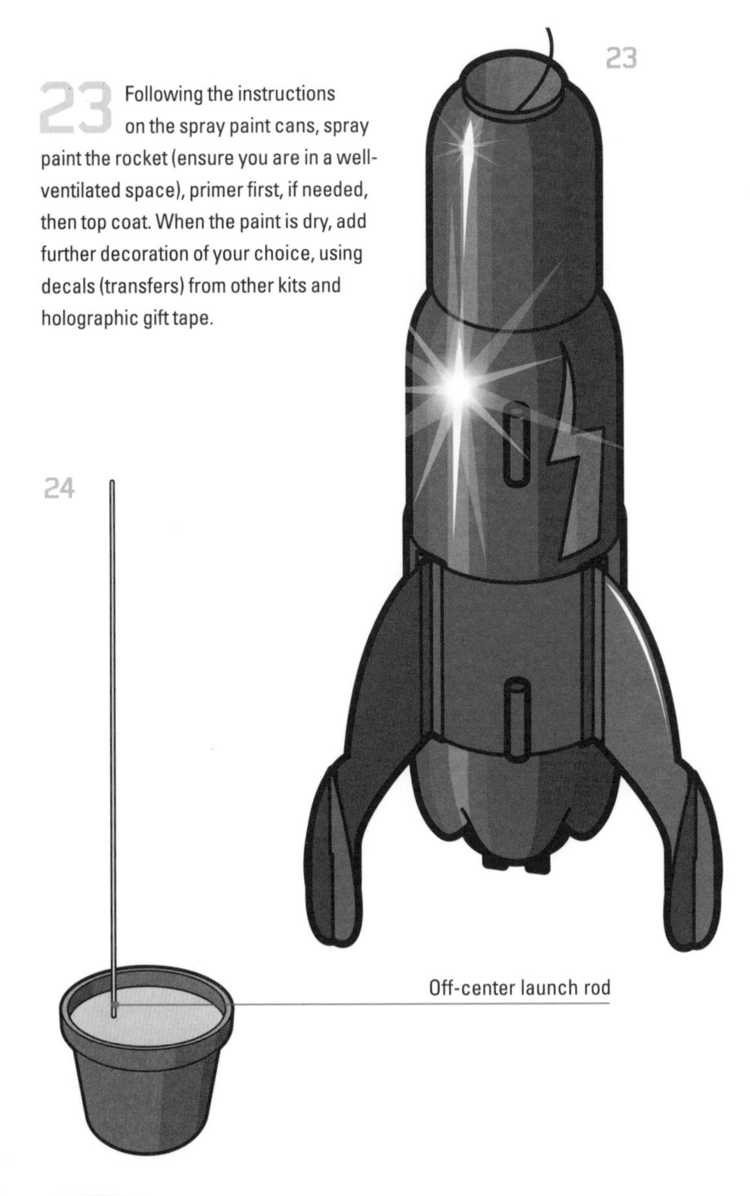

Off-center launch rod

24 To make the launchpad, drill a hole in a block of wood large enough for the 1/16 inch (2mm) metal rod to fit in. Alternatively, fill a small plastic flowerpot with cement (or a mixture of sand and plaster) with the rod secured off-center so the center of the rocket is in the center of the launcher base. Grease the rod so that it can be removed. When the launchpad is not in use, you must cap the upper end of the rod with a cork to avoid eye injuries.

⚠ Safety

- Safety is paramount and there are rules for flying rockets that you must follow. These can be found at the National Association of Rocketry site: www.nar.org/safety.html.
- Consult the safety code and ensure your flying site is suitable. A "D" powered model needs a minimum area of 200 by 200 yards (180 by 180m).
- Remember that the distance the model will travel depends on the size of the parachute and the wind. Do not fly if the wind is above a light breeze.
- The D-12-3 motors (engines) are sold in packages of three and contain detailed instructions on how to insert the igniter, as well as the launch procedure. Be sure to familiarize yourself with these if you have not flown a model rocket before.
- Only fly in an adequate open space and make sure there are no dry materials that can be ignited by the exhaust.
- If possible, place the launcher on a table or stand to keep the rod end above eye level.
- Keep spectators well away; they will be safer and will also get a better view.

For further information and advice on safety considerations, consult the following weblinks:

http://tinyurl.com/dk89vu
http://tinyurl.com/c559b5
http://tinyurl.com/cn6n3c
http://tinyurl.com/c2t9ed
http://tinyurl.com/cb59zo
http://tinyurl.com/cyfuz3
http://tinyurl.com/d89gxy
http://tinyurl.com/dg2o4k

The Rocket

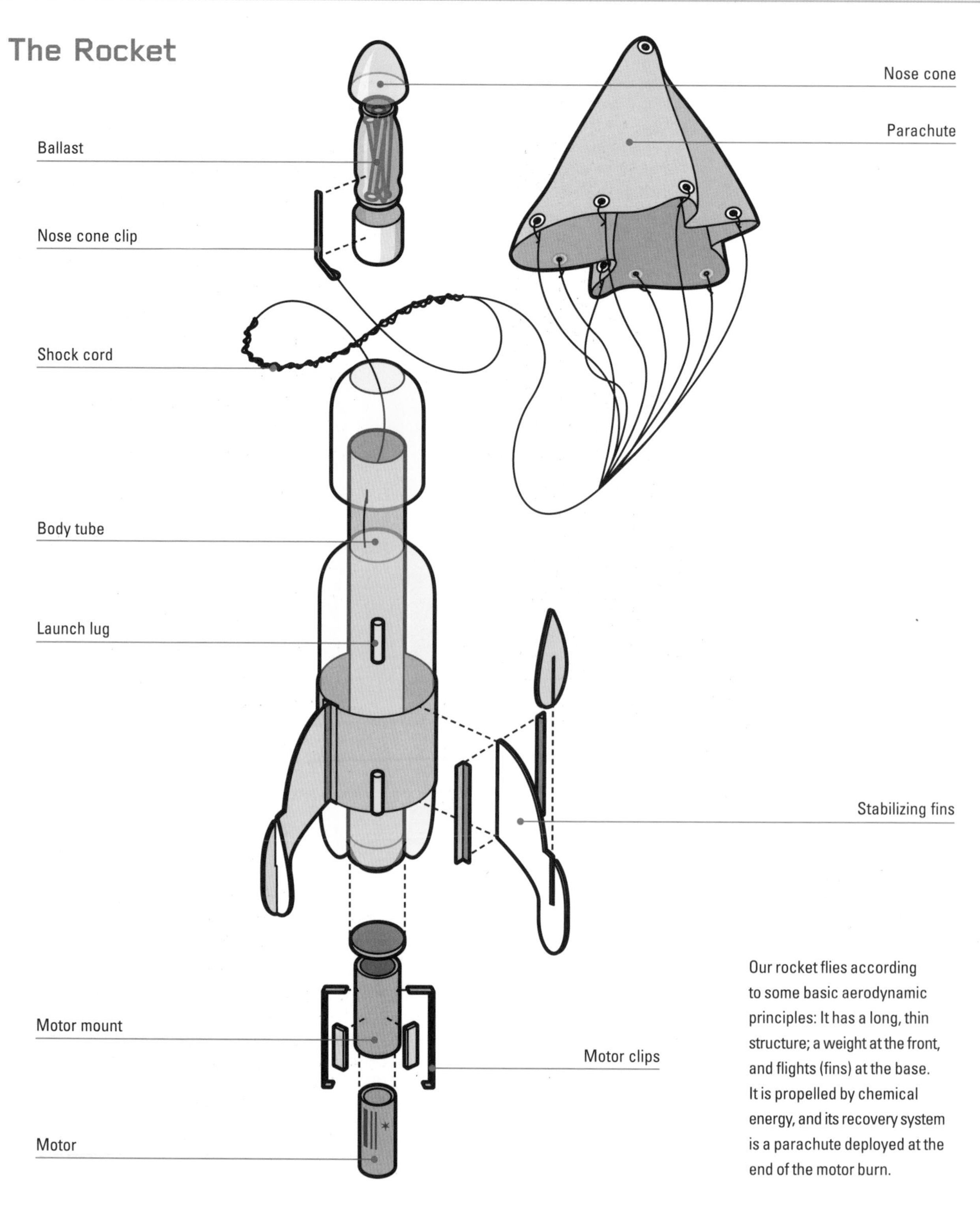

Our rocket flies according to some basic aerodynamic principles: It has a long, thin structure; a weight at the front, and flights (fins) at the base. It is propelled by chemical energy, and its recovery system is a parachute deployed at the end of the motor burn.

How to Use It

Now that you've completed your rocket, you're almost ready for your maiden rocket flight. All you need is some fine weather and an appropriate launch site. Before you start preparing for liftoff, you must read through all the safety points on page 29, and check the National Association of Rocketry's website.

Setting up the Launch Area

To prepare your rocket for launch, first set up your launch pad on a firm surface, such as a table, well away from any spectators. Solder a pair of small alligator clips to one end of the 16 feet (4,880mm) of twin wires. These will be attached to the igniter when you are ready to launch. One strand of the wire is ready to connect to the battery and the other needs a push button between it and the other battery terminal—about a yard from the battery is convenient. To comply with the safety rules, one terminal must be disconnected until you are ready to launch.

Final Preparations to the Rocket

Roll a ball of aluminum foil and push it down from the top of the body tube. Add a couple of squares of foil about 3 inches (80mm) square (these are to protect the parachute during ejection at the end of the flight).

Fold the parachute along its length, then fold in half, tucking the strings into the bottom fold. Push the chute with the shock cord into the body tube and put the nose cone in place. The chute must not be too tightly packed in or it will not eject. Insert the motor into the motor mount at the bottom of the body tube and clip into place.

Countdown to Lift Off!

Follow the instructions that come with the motors for putting the igniter in the motor, and slide the model onto the launch rod. Make sure it can slide freely on the rod. Clip the alligator clips to the igniter and then make the connections to the battery. The model is ready to fly!

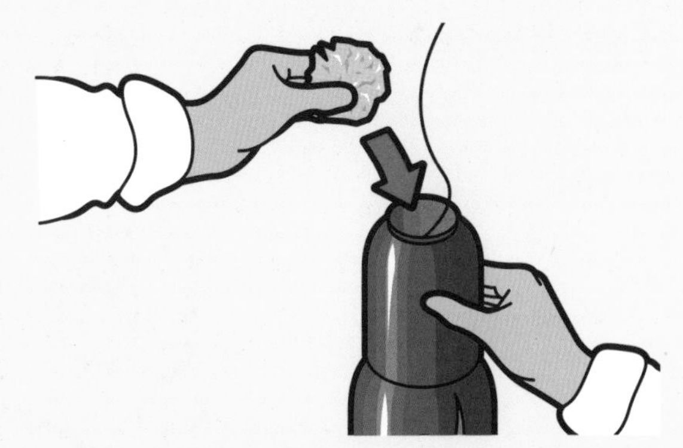

Push a ball of aluminum foil into the rocket body.

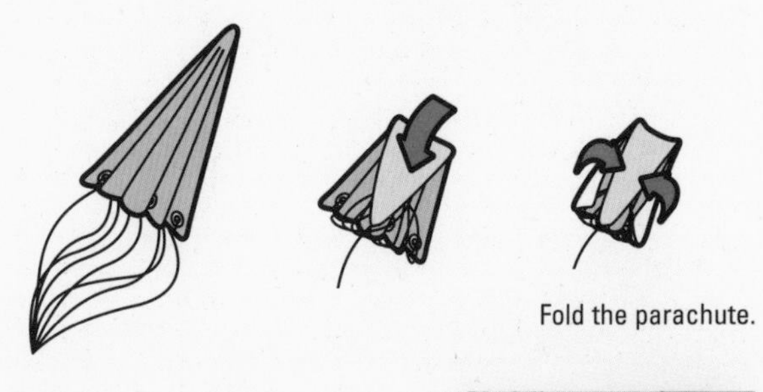

Fold the parachute.

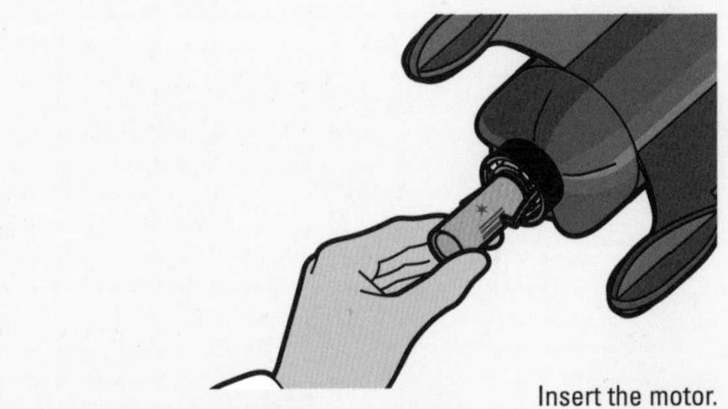

Insert the motor.

The rocket is ready to launch... and land.

Lemon Cannon

As the saying goes, if life gives you lemons... put them into a homemade cannon and see just how far they can go. Our record for slinging a blameless piece of citrus stands at around a hundred yards, but you may be able to improve on this. The cannon fires under pressure; air is pumped into the pipe by a bicycle foot pump, then released by means of a Schrader valve. The key to a successful flight is to ensure that all the joints are rock solid, so they won't give as the pressure builds. Start, too, with a modest amount of air—if you begin with too much, it's more likely your cannon will spit than your lemon will go stratospheric. This is a good project to try out with kids: You don't need too much time or patience and it's exciting but relatively safe in action.

YOU WILL NEED

- 2-inch (50-mm) end cap
- Schrader air valve
- 2-inch (50-mm) straight connector tube
- 6-foot (2,000-mm) length of 2-inch-(50-mm) diameter PVC plastic pipe
- Two 2-inch (50-mm) threaded connectors
- Two 2-inch (50-mm) male-to-¾-inch (20-mm) female threaded reducers
- 20-inch (500-mm) length of ¾-inch-(20-mm) diameter copper pipe
- ¾-inch (20-mm) ball valve
- Bicycle foot pump

TOOLS

- Adjustable wrench
- Awl
- Cordless drill/driver with a good selection of drill bits to fit
- Hacksaw
- Knife
- Pencil
- Pipe cutter
- PVC primer and adhesive
- Sandpaper (medium grade)
- Silicone sealant
- Steel ruler
- Wrenches

How to Make the Lemon Cannon

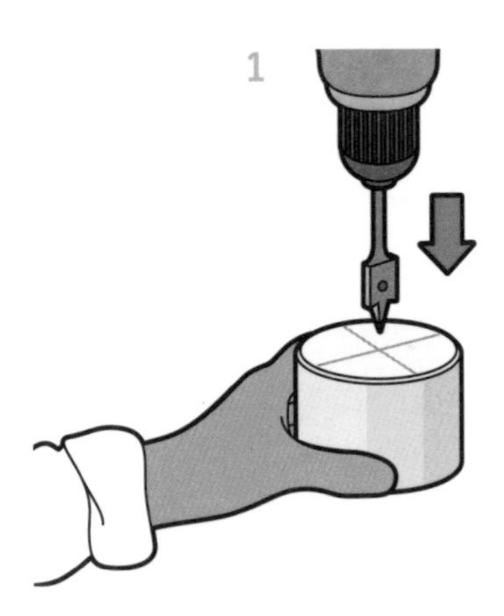

1 Begin by adapting the end cap to take the Schrader air valve. Use a steel ruler to mark a cross at the center of the base of the cap. Drill a ½-inch (12-mm) hole using a spade bit held in a mounted drill. The drill may melt the plastic as it cuts, so clean away any melted plastic before fitting the valve.

2 Use a nut to secure the Schrader air valve to the end cap, making sure that the rubber washer sits firmly inside the cap for an airtight seal. Use a little silicone sealant to be absolutely certain of a hermetic seal.

3 Now you can begin to join the sections of the cannon. First, add the 2-inch (50-mm) straight connector to the 6-foot (2,000-mm) length of PVC plastic tube (the body tube). Clean the contact surfaces of any grease and then evenly coat them with PVC primer. Once the primer is dry, apply the PVC adhesive to both surfaces, then immediately twist the joint to ensure the glued surfaces make contact and let dry.

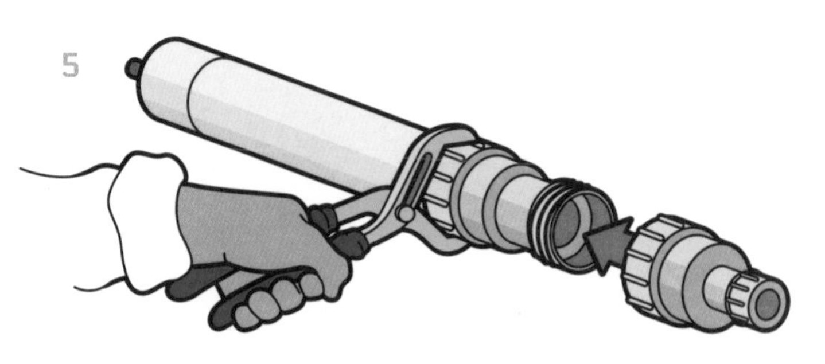

4 Next, attach the straight connector onto the end cap using PVC primer and adhesive, making sure the internal connector is evenly spaced. After 10 minutes, saw the plastic body tube to length. Measure a 2-foot (600-mm) length from the glued joint. Turn the tube against a pencil to mark the cut line, and then cut using a hacksaw. Clean away any rough material with a knife.

5 Now fit the threaded connector to the newly cut end of the body tube to complete the compression chamber. Again, use some silicone sealant to ensure that all mechanical joints are airtight. Next, fit a reducer to the threaded connector, ready to fit the middle section of the lemon cannon—the release tap.

6 The release tap comprises an in-line compression ball valve, with an attached handle. Use the pipe cutter to cut the 20-inch (500-mm) length of copper pipe into two 10-inch (250-mm) lengths. Secure each length of copper pipe on either side of the release tap by using two wrenches turned in opposite directions. The brass balls will compress to make an airtight seal, but add a little silicone sealant to make sure.

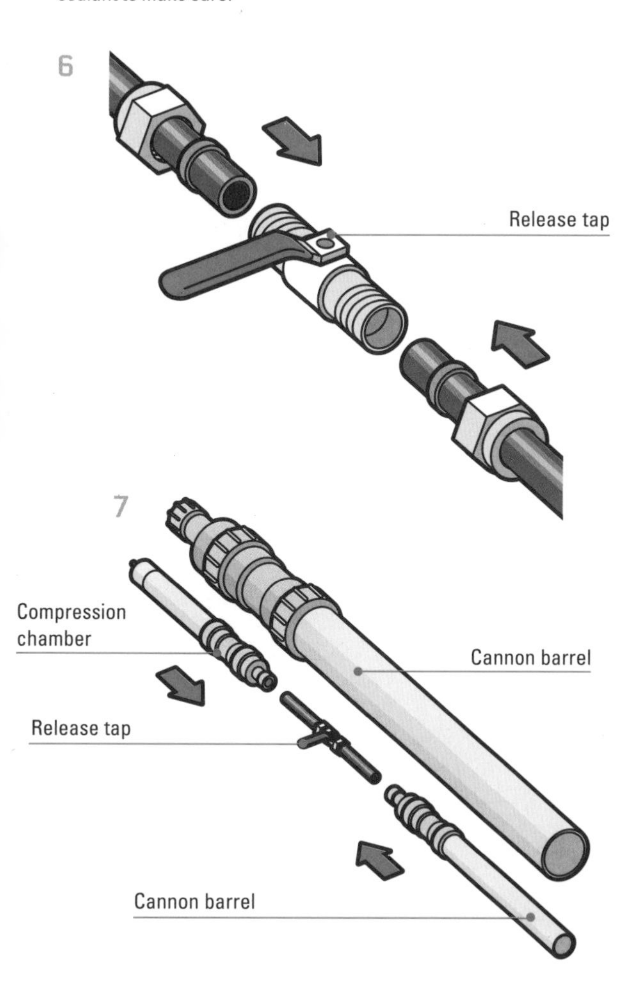

7 The barrel of the cannon is a mirror image of the compression chamber, except that it isn't capped at the end. Instead it uses up the remaining 4-foot (1,200-mm) length of PVC plastic tube. Now that you have the compression chamber, the release mechanism, and the barrel complete, simply join them all together by using the two plastic compression joints at the end of the chamber and the barrel.

How to Use It

Close the compression valve, and connect the bicycle foot pump to the Schrader air valve and pump it cautiously until it becomes difficult to push. Don't exceed the pipe rating; most pipes will take 35 psi (pounds per square inch). As you pump, see if air is coming out of any of the joints (you can loosely cover lengths of piping with plastic wrap; you'll see it move if there are any leaks). If you find any leaks, seal again with PVC adhesive. Hold the cannon at a 45-degree angle to achieve maximum trajectory. Finally, select a lemon, or similar fruit, that fits into the cannon's mouth. Check that the cannon is not pointing at anyone and open the valve with one swift action.

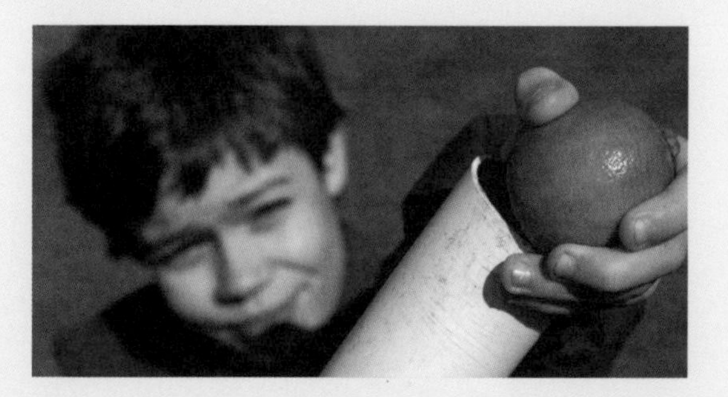

Tip

Use a fruit or vegetable that fits into the barrel well, but not tightly. If it is wedged in or too small, the compressed air will escape without propelling the projectile. Lemons and limes are a good choice because they are round and elongated. If the fruit is too large, try trimming some peel off. If it is too small, try adding duct tape to increase its girth.

Heli-Whirler

This project looks great in flight: A spinning, silvery disk whirling through the air at high speed. The "spin" is simply created by pulling a cord wound around the wooden reel at the top of the handle: the reel turns fast, and the rotational force causes the disk above it to lift into the air—and to fly an impressive distance. It's likely that you'll be won over by the heli-whirler's elegant simplicity.

A word of warning: Make sure that you don't omit the glued edge around the raw metal rim, and be very careful where you choose to launch your whirler. The edges are sharp and the flight fast, so fly responsibly and stay safe.

YOU WILL NEED

- Sheet of thin tinplate, about 7 × 7 inches (180 × 180mm)—we used the lid of a metal cookie box
- Galvanized coat hanger (or a length of wire)
- 12-inch (300-mm) length of 1-inch- (25-mm) diameter dowel (this allows for cutting waste)—we used a wooden broom handle
- Reel for holding thread—wood is best but plastic will do
- Two 1¼-inch- (30-mm) diameter metal washers
- Two finishing nails and a screw

TOOLS

- Cordless drill/driver with a good selection of drill bits to fit
- Craft knife
- Gas torch for soldering
- Goggles for soldering
- Pair of dividers
- Pair of tin snips
- Pencil and ruler
- Pliers
- Roll of flux-core solder
- Sandpaper (fine grade) and sanding block
- Screwdriver
- Small file
- Small hammer
- Staples for holding the wire prior to soldering
- Straight handsaw
- White glue
- Workbench with vise

How to Make the Heli-Whirler

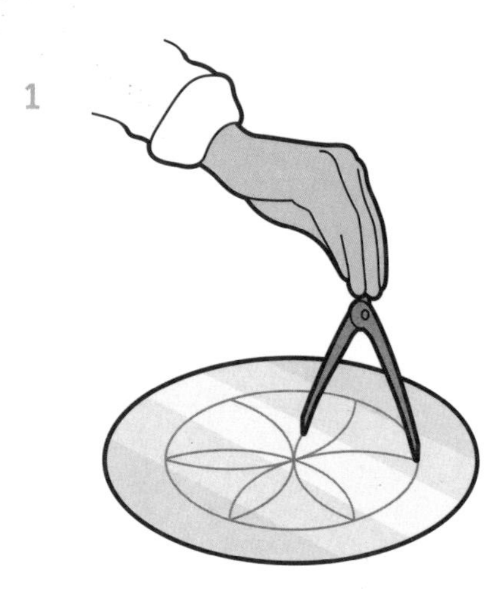

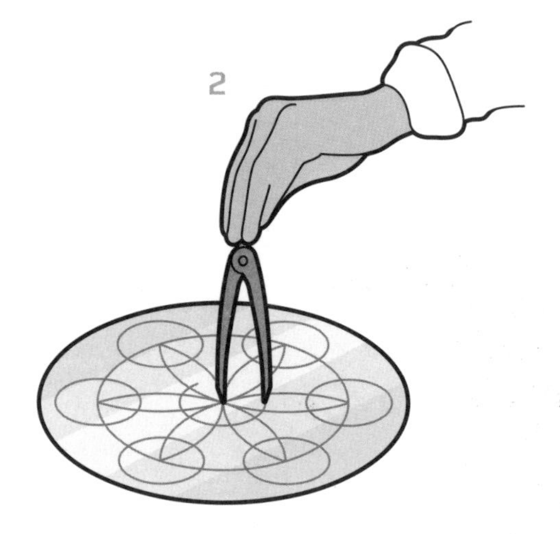

1 Set the dividers to a 3-inch (75-mm) radius and scribe out a circle 6 inches (150mm) in diameter on the tinplate. With the dividers set to the same radius, spike the point on the circumference of the circle and scribe an arc that intersects the circumference at two points. Continue working around the circumference, spiking the intersections and drawing arcs until you have a six-petal "hex-flower" motif.

2 Set the dividers to a radius of 1 inch (25mm), spike it on a petal point, and draw a circle 2 inches (50mm) in diameter. Do the same with each of the other petal points, and draw another circle at the center of the petal motif.

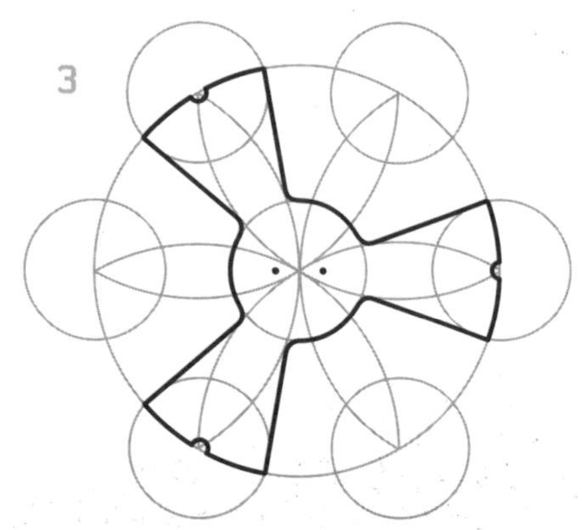

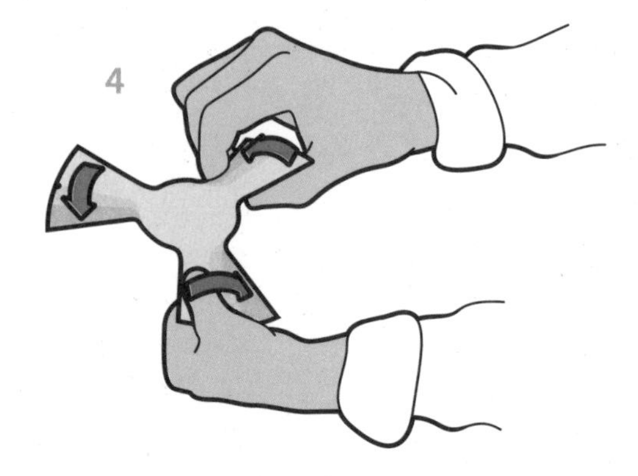

3 Take a ruler and dividers and, using the underlying hexagon as a guide, set the dividers to a radius of 1 inch (25mm) and draw six 2-inch- (50-mm) diameter circles, one at the center of the flower, and one at each petal point. Next, use a pencil and ruler to draw in the three-blade whirler shape, as illustrated.

4 Use the tin snips to cut out the three-blade shape, complete with end-of-blade notches. Be careful: The cut edge of the tin is razor sharp! You might need to wear heavy-duty gloves. Use the file to remove the burr from the cut edge of the tin. Twist the ends of the heli-whirler blades slightly so that when each blade is seen from the end, it angles from top left down to bottom right.

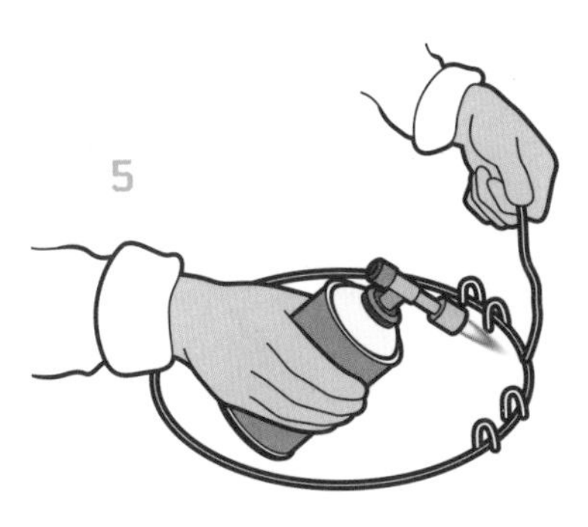

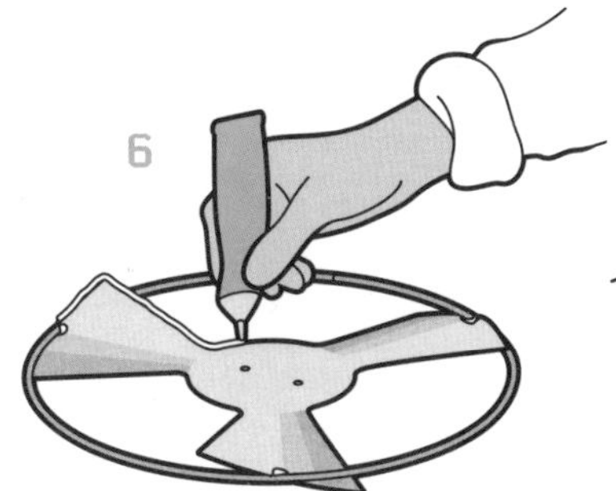

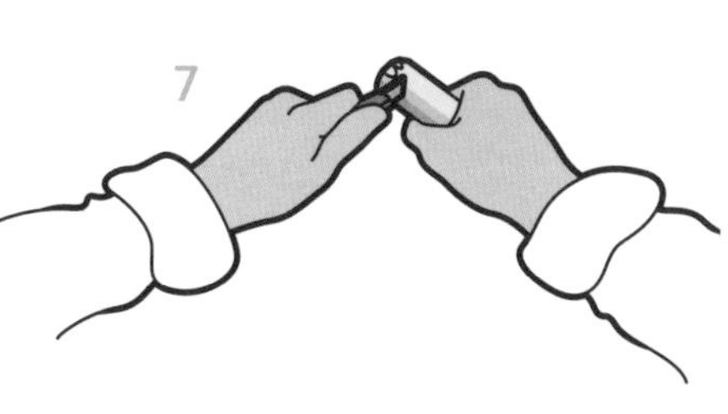

5 Take the metal coat hanger and use your hands and pliers to make a 6-inch (150-mm) hoop. Clean the joining ends of the wire with the file (to remove the zinc) and use the gas torch and solder to join the ends. Remember to wear goggles while soldering and clip the hoop with staples to hold it still.

6 When you have made the hoop, set the three-blade, tin whirler in place (so that the notches at the end of the blades are positioned on the wire) and then solder it in place. Note how, from one blade to another, the "angle of tip" is identical. Dribble a bead of white glue all around the edge. (Note that, although this will buffer the edge, the edge will still potentially be very sharp if wear and tear cause the glue to peel off.)

7 Using the handsaw, cut the dowel down to a length of 7½ inches (190mm). Use the craft knife and sandpaper to shape the ends so that one end is rounded and the other flat and smooth.

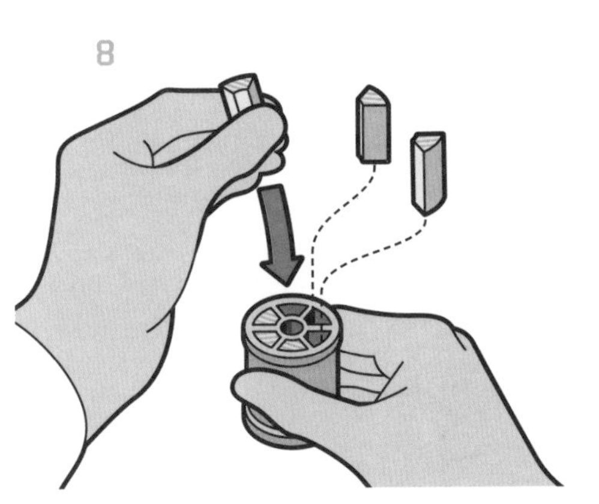

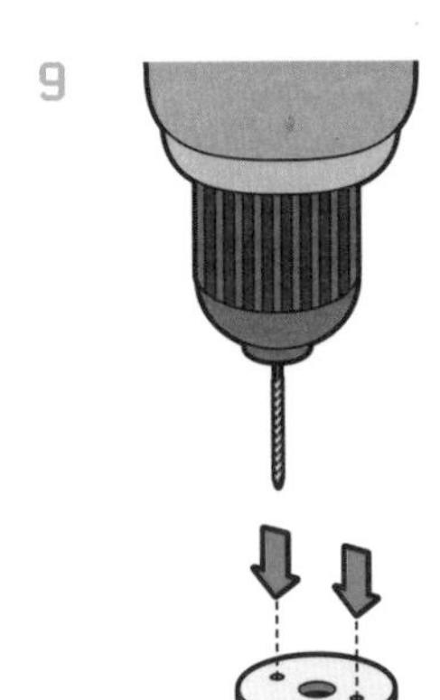

8 If you have a traditional, wooden thread reel, you can use it as it is. If it is plastic, as here, fill the cavities with scraps from the dowel waste (trimmed with the craft knife to a tight fit) and rub them down to a smooth finish.

9 Take one of the two metal washers, mark location points about ¾ inch (20mm) apart (each about ⅜ inch [10mm] out from the center) and drill them through with a ⅛-inch- (3-mm) diameter drill bit.

⚠ Safety

- The edges of the tinplate, once cut, are razor sharp, so pay special attention while handling it, and run a bead of white glue around the edge, as indicated.
- Be careful when paring the wooden dowel with the craft knife.
- Goggles are advised when soldering and when flying the machine.
- Warn children to be careful when handling the heli-whirler.

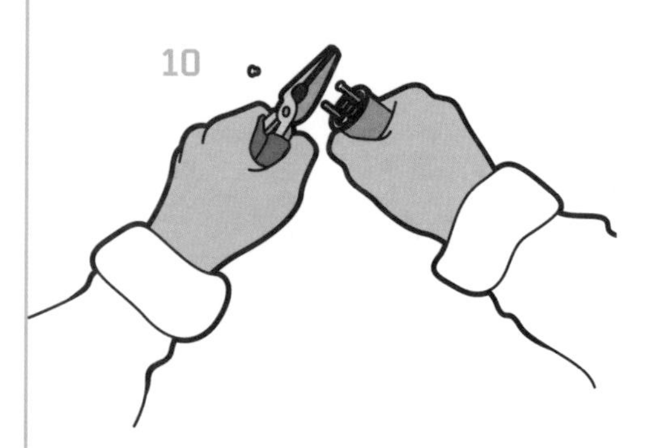

The Heli-Whirler

10 Center the drilled washer on the thread reel, and use the hammer to tap two nails firmly into place through the holes. Use the pliers to snip off the top of the nails—so that you are left with two ½-inch- (12mm) long location brads.

11 Set the washers and thread reel on top of the wooden handle (undrilled washer first, reel next, then washer with location pins uppermost) and attach them in place with a screw running down into the top of the handle.

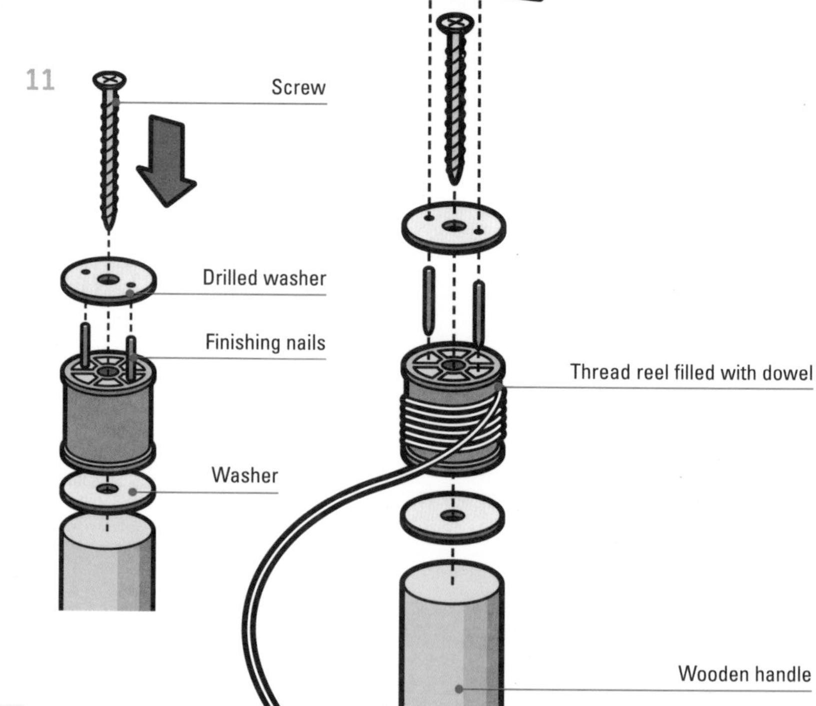

How to Use It

First, set the heli-whirler on top of the nails. The heli-whirler is then operated by winding a length of string clockwise around the thread reel, and then pulling it hard, so that the reel spins rapidly under its own momentum... ZOOOOOM!

Rotational force causes our heli-whirler to spin and lift. The angle of the blades, the weight of the wire, your soldering skills, and the direction that you wind the cord around the reel all have an impact on the flight's success.

MACHINES THAT SWIM AND FLY

Indonesian Fire Kite

If you have ever played that after-dinner game in which you roll candy papers into loose little cylinders, set them alight, and watch them float up to the ceiling—this is the grown-up and much enlarged version. The paper kite is neatly folded, set on fire, and, on a still day, will float skywards to a surprisingly great height.

This is the simplest project you'll find in our collection, but it requires careful making and lighting, and it may take a few tries before you get a perfect flight. The "kite" effect depends on the lit corners of the kite heating the pockets of air trapped within the paper. Surrounded by cooler air, and operating on the principle that heat rises, the kite will rise up and fly. The more evenly it burns the better the resulting flight, so it may help to get a few friends and some gas-flame lighters or fireplace matches, and light all four corners at exactly the same moment.

YOU WILL NEED

- Sheet (two-page spread) of a large-size newspaper
- Adhesive tape

TOOLS

- Four barbecue gas-flame lighters, or four long-handed fireplace matches
- Scissors

How to Make the Indonesian Fire Kite

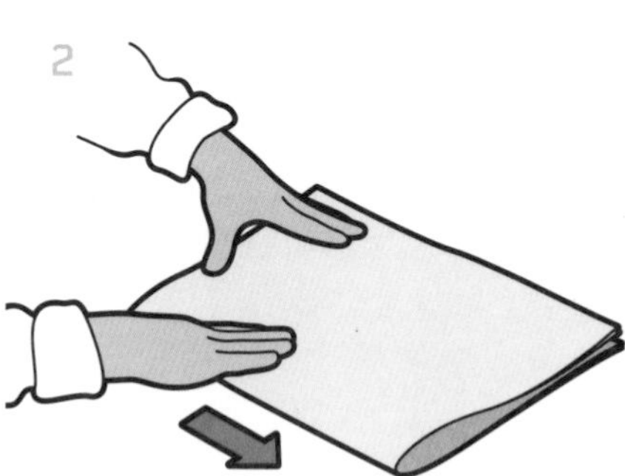

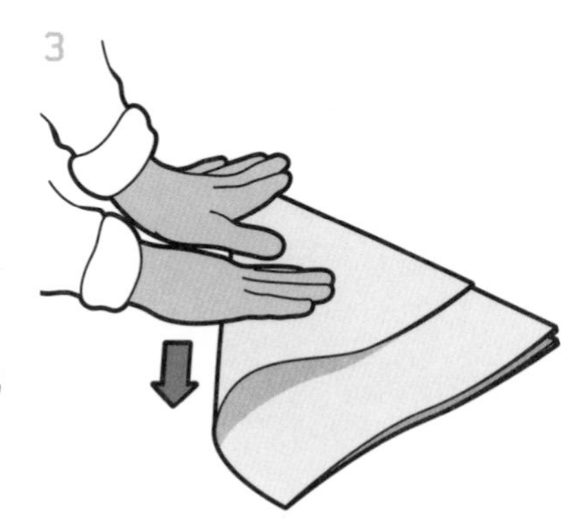

1 Fold the sheet of newspaper exactly in half, back along its original crease but pressing firmly to make this sharper.

2 Fold the bottom half upward as shown and smooth down the crease.

3 Take hold of the bottom left-hand corner and fold it upward toward the top edge, making a 90-degree triangle. Align the top edges accurately and hold them in alignment, at the same time smoothing down the 45-degree crease.

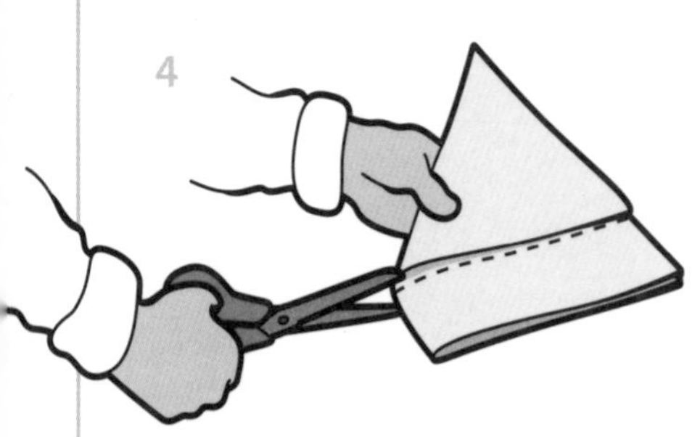

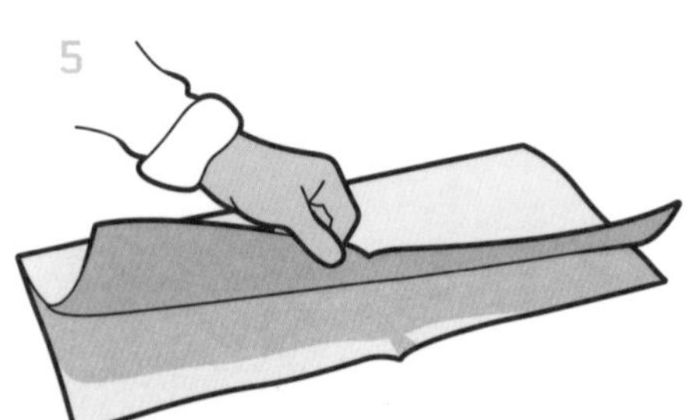

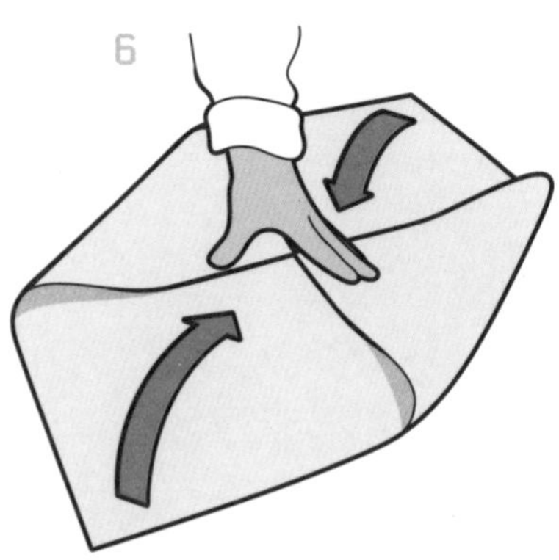

4 Using a sharp pair of scissors, carefully cut off the waste on the right-hand side.

5 Unfold the newspaper and smooth it down to make a flat square.

6 Fold all four corners of the paper toward the center, so that they meet in the center as accurately as possible. Do not press down on the folds; you are making an "air-filled cushion"—not a flat envelope.

7 Use a single length of adhesive tape about 1 inch (25mm) long to join the corners to each other. Keeping the weight of the kite to a minimum is important, so do not use more tape. If you do, the kite will take longer to rise. Avoid sticking the corners to the paper beneath.

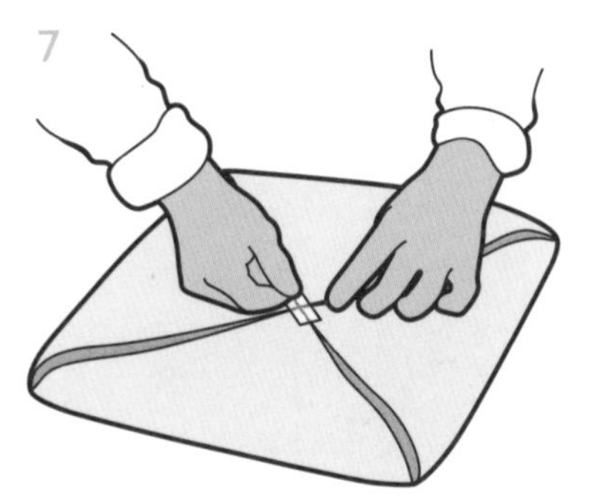

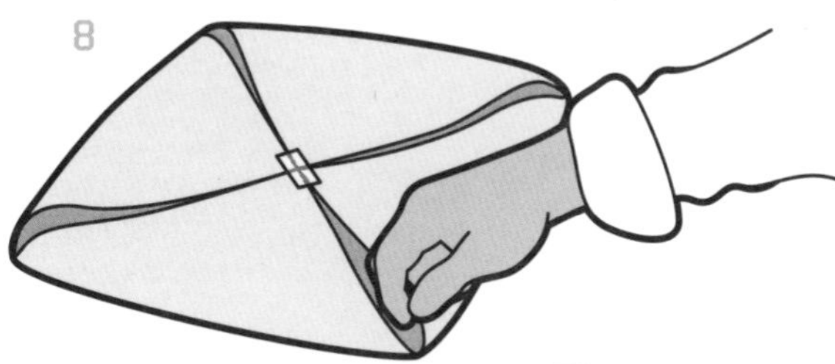

8 Tease out any of the new corners if they are a bit crumpled and make sure that the whole shape resembles a square plumped cushion cover.

⚠ Safety

- Exercise caution when lighting the kite to avoid burns.
- The kite is lit while on the ground, making it difficult to use a lighter without burning your fingers. To avoid this, use four barbecue gas-flame lighters or long-handed fireplace matches.
- The ideal site for releasing the kite is a remote location, where there are no people, buildings, or overhanging structures. Ensure there is nothing flammable nearby, both on the ground (such as paper or dry grass) or above ground (trees, utility poles, and so on).

How to Use It

The spectacular sight of your blazing kite rising into the sky is moments away. For obvious reasons, a still day (or, for the most romantic kite-flying, a still evening) is the best time for its maiden—and only—flight.

Lighting the Kite

Take the kite to an outdoor area suitable for launching. Place it at ground level on a noncombustible surface, paper seams facing downward. Ideally, you need three assistants, each with a lighter to ignite all corners of the kite simultaneously. If one corner burns too quickly, the kite will not fly. One person can light all corners quickly using a lighter in each hand, but he or she would need to move swiftly.

The kite will burn toward the center and the inside will fill with hot air, making the kite buoyant. As the paper burns, the kite gets lighter in weight, becoming a glowing ember that rises higher until the paper disintegrates.

Paddle-Operated Punt

There's a retro quality to this wooden punt, perhaps because it was inspired by an example we found in a boy's woodworking periodical from the 1910s. We've adapted it—the original was too elaborate and time-consuming in its construction for today's busy maker—but you still need reasonable woodworking skills and some patience to make it, so don't choose it as your very first woodworking project (try an easy project, such as the pneumatic boat on pp. 58—61, if you're just starting out). Driven by foot-operated paddles, which turn two wheels, one on each side, that churn it through the water, the finished punt is fun to watch as well as to operate.

YOU WILL NEED

Pine to make up:

- A: piece, 108 × 15 × 1¾ inches (2,750 × 380 × 43mm)
- B: piece, 15 × 2¾ × 1¼ inches (380 × 70 × 32mm)
- C: piece, 35½ × 10 × 1¾ inches (900 × 255 × 43mm)
- D: two pieces, 6 × 3½ × 3½ inches (150 × 90 × 90mm)
- E: two pieces,19 × 1¼ × 1¼ inches (500 × 32 × 32mm)
- F: two pieces, 63½× 1¼ × 1¼ inches (1,610 × 32 × 32mm)

Plywood to make up:

- G: two pieces, 60 × 11 × ¾ inches (1,525 × 280 × 18mm)
- H: piece, 18× 10 × ¾inches (460 × 255 × 18mm)
- I: two pieces, 30 × 8 × ¾ inches (760 × 200 × 18mm)
- J: eight pieces, 12 × 6 × ½ inches (300 × 150 × 9mm)
- K: piece, 18 × 9 ¾ × ½ inches (460 × 250 × 9mm)

Steel to make up:

- L: two pieces, 11 × 1¼ × ⅛ inches (280 × 32 × 4mm)
- M: two pieces, 6¾ × 2¼ × ⅛ inches (170 × 56 × 4mm)
- N: two pieces, 8 × ½ × ⅛ inches (200 × 14 × 4mm)
- O: piece, 8¾ × 2¼ × ⅛ inches (220 × 56mm × 4mm)
- One 3½-inch (90-mm) carriage bolt, one big washer, one small washer, and two nuts (to join foot-operated lever to hull)
- Fourteen 1¼-inch (30-mm) no. 10 screws (for laminate outrigger)
- Five 3-inch (80-mm) no. 10 screws (to join seat support to hull)
- Four 4¼-inch (110-mm) bolts, eight washers, and four nuts (to join floats to outrigger)
- 4-inch (100-mm) no. 10 screw (to attach outrigger to seat support)
- Four 2-inch (50-mm) no. 10 screws (to join seat to seat support)
- 39-inch (990-mm) length of ⅜-inch- (10-mm) diameter threaded rod (for crankshaft)
- Four 2½-inch (60-mm) bolts, four nuts, and four washers (to join crank brackets to hull)
- Thirty-two 1½-inch (40-mm) no. 10 screws (to join paddles to paddle wheel blocks)
- Six ⅜-inch (10-mm) nuts, two small washers, two big washers, two 2-inch (50-mm) square washers
- Four 1-inch (25-mm) no. 10 screws (to join paddle wheels to crank)
- 5½-inch (140-mm) length of 5/16-inch- (8-mm) diameter threaded rod, four nuts, two big washers, and two small washers (to join hand levers to seat support)
- Four 2-inch (50-mm) bolts, ¼ inch (6mm) in diameter, and two nuts (to join loops to connecting rods)
- Two 3-inch (80-mm) bolts, 2⅜ inches (6mm) in diameter, four nuts, two big washers, and four small washers (to join hand levers to connecting rods)
- 14¾-inch- (350-mm) length of ⅜-inch- (10-mm) diameter threaded rod (to run from rudder)
- Four 1⅜-inch (35-mm) bolts, ¼ inch (6mm) in diameter, and 4 nuts (to join loops to rudder)
- Two ⅜-inch (10-mm) nuts, two big washers, and one small washer (to join rudder to rudder lever)
- Two 62-inch- (1,580-mm) lengths of ⅛-inch- (3-mm) diameter wire rope, plus wire rope clips
- ⅜-inch (10-mm) dome nut (for rudder rod)
- Four 1-inch (25-mm) no. 10 screws and four washers (to join wire rope to levers)
- Marine varnish
- Sixteen 19½-inch (500-mm) lengths of ¾-inch- (20-mm) wide webbing
- Sixteen 1-inch (25-mm) no. 10 screws (to join webbing to floats)
- Two 26-inch (660-mm) bicycle inner tubes
- Foam insulation board, 8 feet × 4 feet × 4 inches (2,240 × 1,220 × 100mm)
- Silicone sealant

How to Make the Paddle-Operated Punt

TOOLS

- Block plane
- C clamp
- Center punch
- Compass (geometrical)
- Compound miter saw
- Cordless drill/driver with a good selection of drill bits to fit
- Drill press (column drill)
- Hacksaw
- Hammer
- Handsaw
- Metal file
- Pencil
- Protective gloves (for welding)
- Rulers, 12-inch (300-mm) and 40-inch (1,000-mm)
- Saber saw
- Sandpaper and sanding block
- Shielded metal arc welder
- Spokeshave
- Spring clips
- Tape measure
- T-square
- Trestle and vise
- Welding helmet
- White glue
- Workbench
- Wrenches

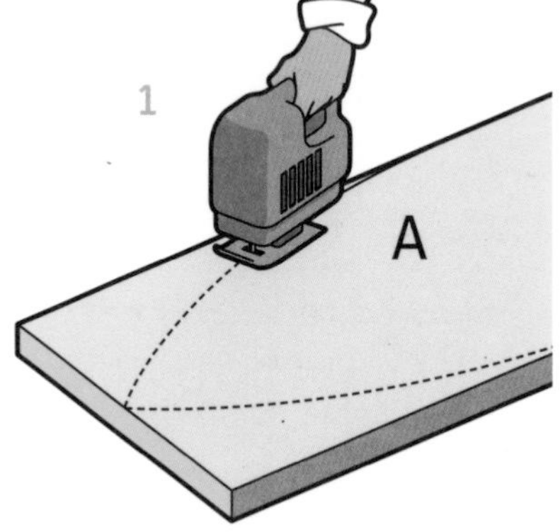

1 Take pine piece A and, using a saber saw, cut out the shape of the hull, curving to a point at the front and rounded at the back. The shape is not critical but should be symmetrical. Round the edges with a block plane and sandpaper.

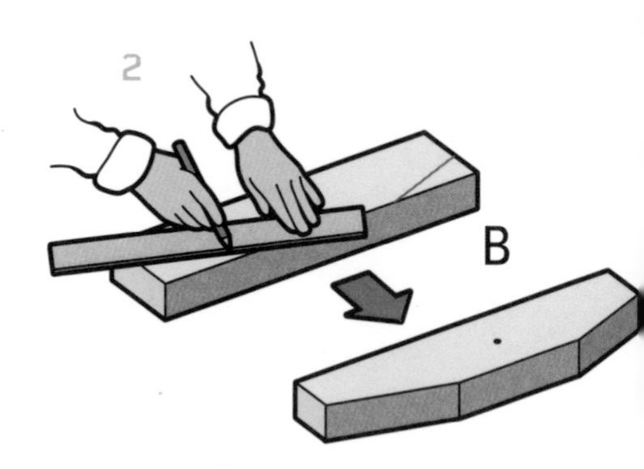

2 To make the foot-operated lever for controlling the rudder, use a ruler to mark out a tapered shape on pine piece B, starting $4\frac{1}{4}$ inches (110mm) from one end and tapering to $1\frac{1}{2}$ inches (30mm) wide at the other (see template on page 153). Cut off the waste with a saber saw and use a block plane and sandpaper to smooth the edges and corners. Drill a $\frac{1}{4}$-inch- (8-mm) diameter hole in the center.

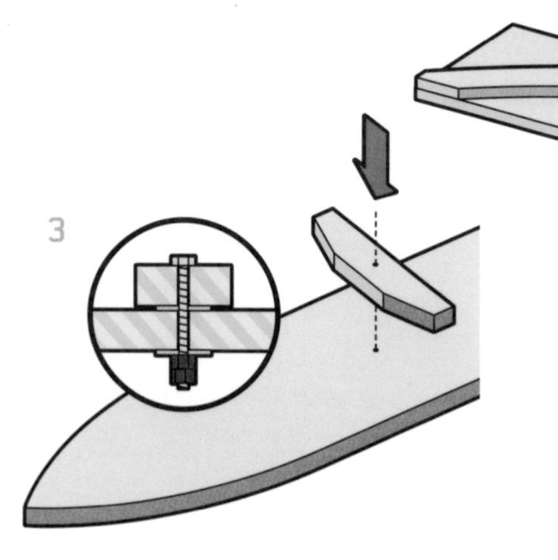

3 Bolt the lever using a $3\frac{1}{2}$-inch (90-mm) long, $\frac{3}{8}$-inch- (8-mm) diameter carriage bolt to the hull, 46 inches (1,170mm) from the front. Insert a large washer between the lever and the hull, and a washer and nuts underneath.

4 On plywood piece G, mark out the shape of the outrigger, with flat ends 4 inches (100mm) long. The width of the curved shape should be about $3\frac{1}{2}$ inches (90mm) wide and have a 10-inch- (255-mm) long central flat area at the top (see template on page 153). To mark out a smooth symmetrical shape, use a flexible plywood cut-off, bending it to the appropriate extent by pushing it against a screw driven into the wood. Cut out the shape with a saber saw (and use this shape as a template to draw around for the second piece. Cut out this piece in turn).

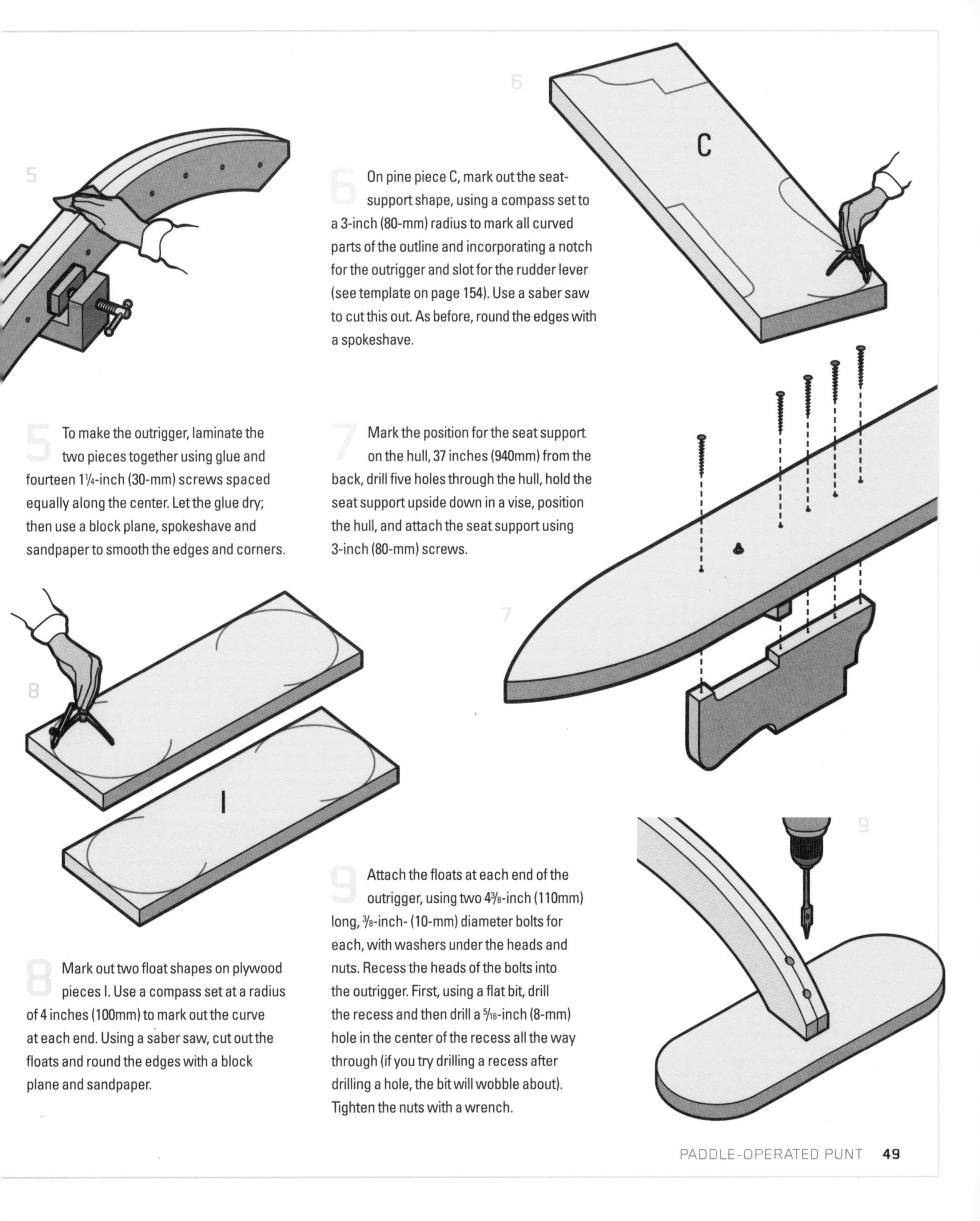

5 To make the outrigger, laminate the two pieces together using glue and fourteen 1¼-inch (30-mm) screws spaced equally along the center. Let the glue dry; then use a block plane, spokeshave and sandpaper to smooth the edges and corners.

6 On pine piece C, mark out the seat-support shape, using a compass set to a 3-inch (80-mm) radius to mark all curved parts of the outline and incorporating a notch for the outrigger and slot for the rudder lever (see template on page 154). Use a saber saw to cut this out. As before, round the edges with a spokeshave.

7 Mark the position for the seat support on the hull, 37 inches (940mm) from the back, drill five holes through the hull, hold the seat support upside down in a vise, position the hull, and attach the seat support using 3-inch (80-mm) screws.

8 Mark out two float shapes on plywood pieces I. Use a compass set at a radius of 4 inches (100mm) to mark out the curve at each end. Using a saber saw, cut out the floats and round the edges with a block plane and sandpaper.

9 Attach the floats at each end of the outrigger, using two 4⅜-inch (110mm) long, ⅜-inch- (10-mm) diameter bolts for each, with washers under the heads and nuts. Recess the heads of the bolts into the outrigger. First, using a flat bit, drill the recess and then drill a 5⁄16-inch (8-mm) hole in the center of the recess all the way through (if you try drilling a recess after drilling a hole, the bit will wobble about). Tighten the nuts with a wrench.

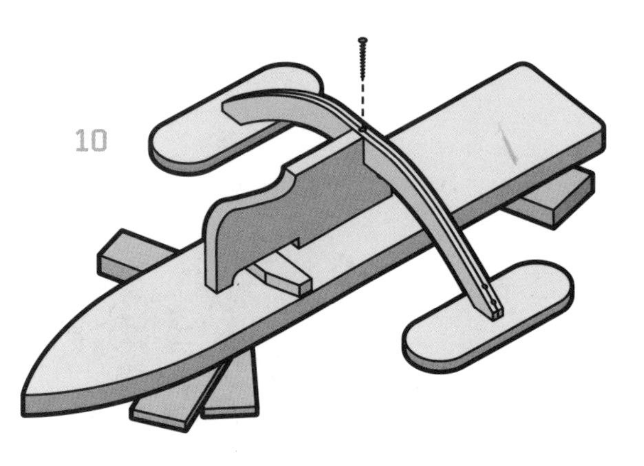

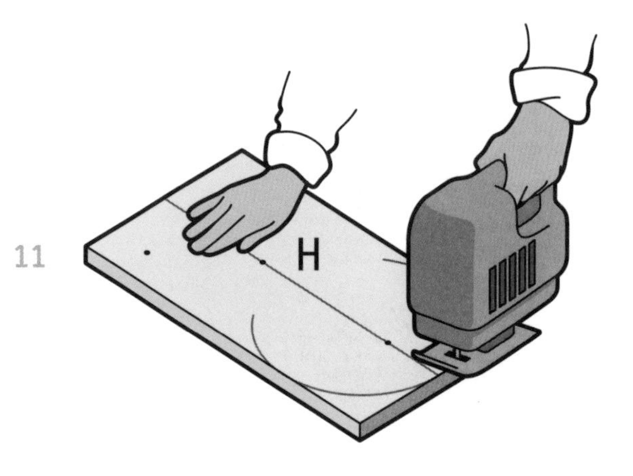

10 Again, using the flat bit, drill a 2-inch (50-mm) recess in the center of the outrigger, then drill a 4-inch (100-mm) hole. Use the 4-inch (100-mm) screw, set in the recess, to attach the outrigger to the seat support. Prop up the hull with scraps of wood while you work.

11 On plywood piece H, mark out the seat shape, using a compass set to a radius of 5 inches (127mm) to mark the rounded front. Draw a centerline and mark one screw hole on this 6¼ inches (160mm) from the back. Mark another hole 3⅜ inches (85mm) from the front. Mark two more holes in the back corners, about 1¾ inches (45mm) in from the sides and 2¼ inches (55mm) from the back edge. Using the saber saw, cut out the seat. Sand the edges of the seat smooth.

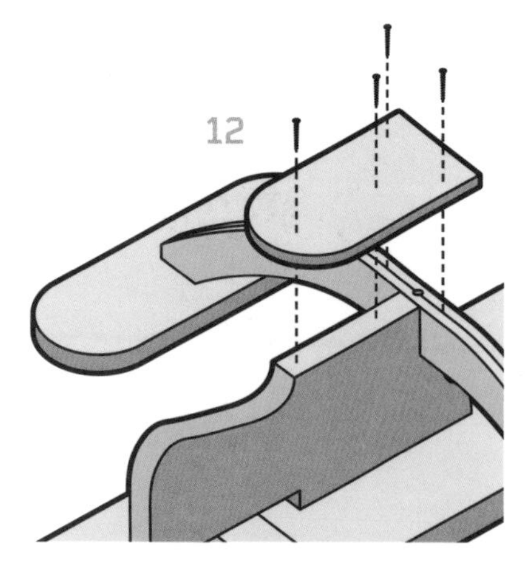

12 Drill 2-inch (50-mm) holes where marked for the screws. Attach the seat to the seat support and outrigger using four 2-inch (50-mm) screws.

13 Take steel pieces L (used for the crank brackets), and file the ends until they are smooth and rounded.

14

14 Using a center punch and hammer, mark the location for holes at the following positions: (a) 1 inch (25mm) from the top end (this is for the crank to pass through); (b) 1 inch (25mm) in from the bottom end; and (c) 3⅜ inches (85mm) from the bottom (these are for bolting the crank brackets to the hull). Using the drill press (a strong adult could use a cordless drill but a drill press is much easier) and a pair of spring clips to hold the metal, drill a ⅜-inch (10-mm) hole at location (a), a ¼-inch (8-mm) hole at location (b) and another ¼-inch (8-mm) hole at location (c). Marking the location first stops the drill bit from moving about.

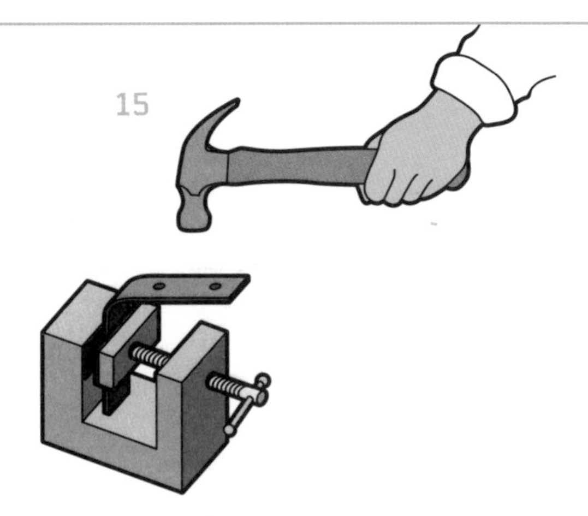

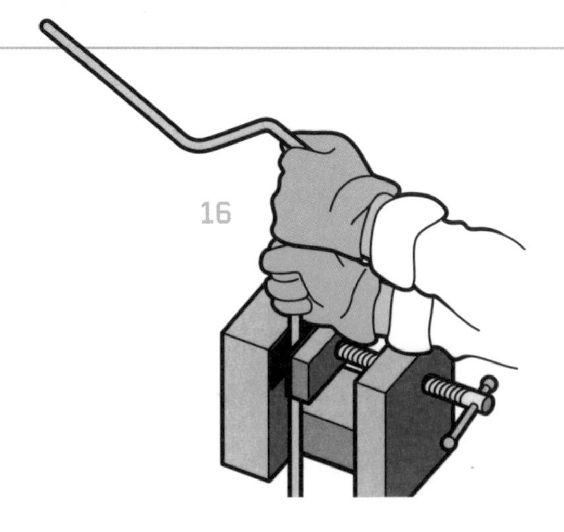

15 One at a time, hold the crank brackets in a vise and bend them 90 degrees at the halfway point. Start by bending by hand and when it gets too difficult to bend farther, use a hammer to beat the metal over 90 degrees.

16 To make the crankshaft, refer to the template on page 154, and bend the 39-inch- (990-mm) length of threaded rod as shown (the angles are 117 degrees) in a vise (wear gloves for this). Proceed carefully: Threaded rod will break if you attempt sharp bends, so push the rod away from your face to avoid any risk of injury.

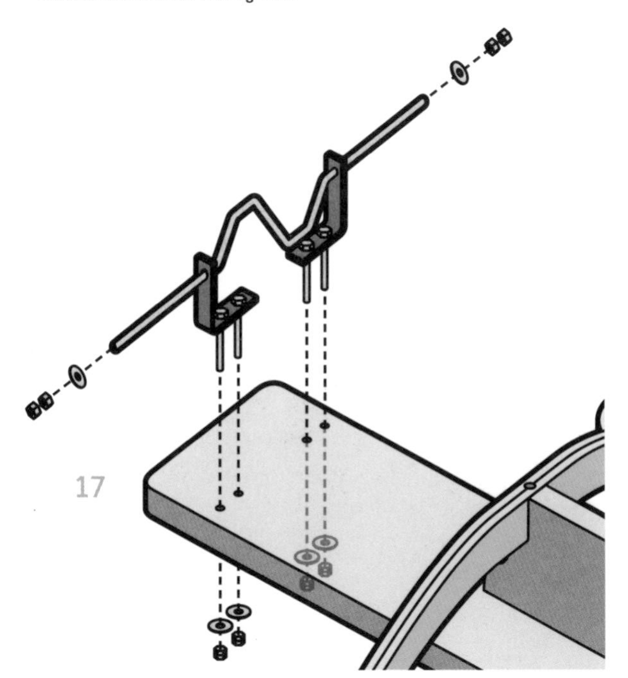

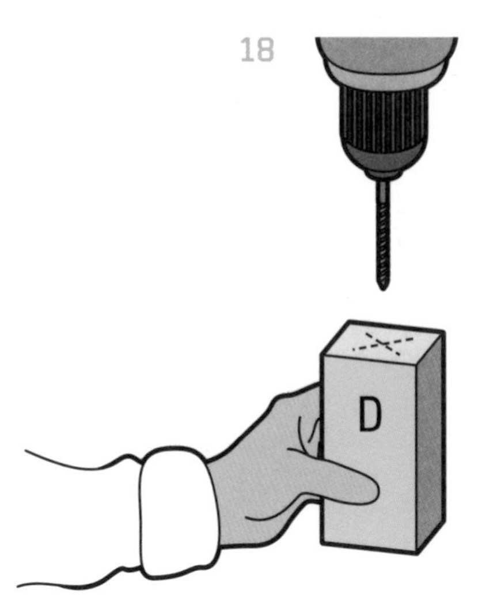

17 Fit the crankshaft into the crank brackets and bolt the brackets to the hull, 6¾ inches (170mm) from the rear end, using two 2⅜-inch (60-mm) bolts (two bolts per bracket). Onto each end of the crankshaft fit a washer and nuts. The crankshaft is held central but rotates freely.

18 To prepare the paddle wheel blocks, mark the center on both ends of pine pieces D, and drill a ⅜-inch- (10-mm) diameter hole through each block, working from both ends (the drill bit will not be long enough to go through from only one side).

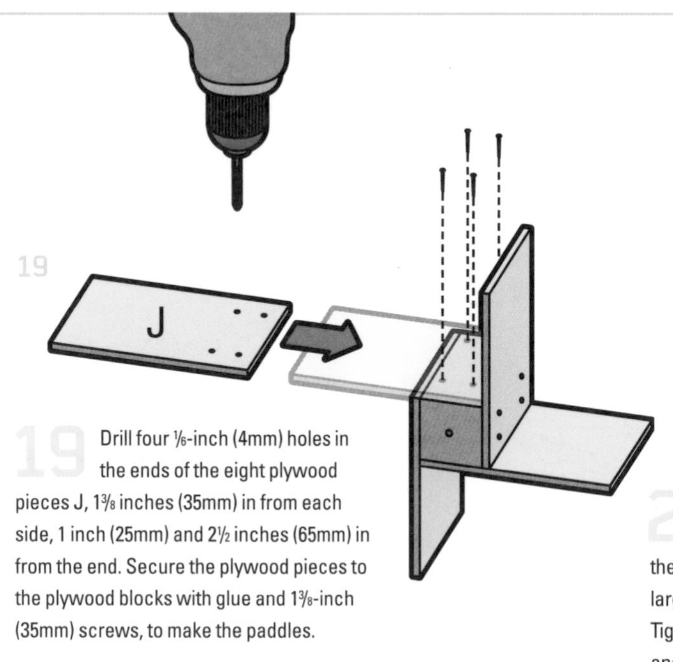

19 Drill four ⅙-inch (4mm) holes in the ends of the eight plywood pieces J, 1⅜ inches (35mm) in from each side, 1 inch (25mm) and 2½ inches (65mm) in from the end. Secure the plywood pieces to the plywood blocks with glue and 1⅜-inch (35mm) screws, to make the paddles.

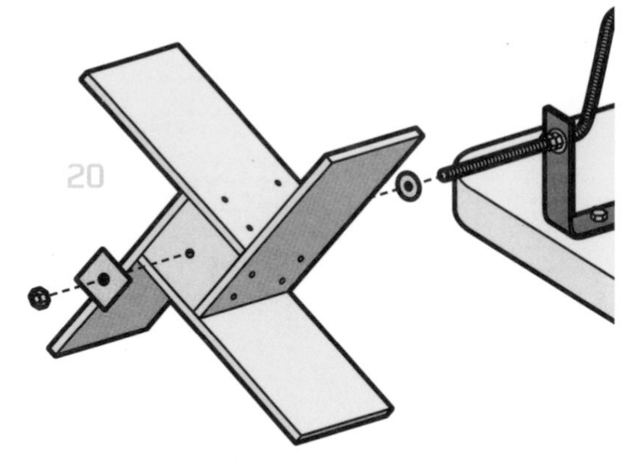

20 Put a large washer on each end of the crank (against the nuts already there), then put a paddle wheel on each end of the crank, inserting the crank through the hole drilled in the block. Put a large square washer after the paddle wheel and then screw on a nut. Tighten the nut so that the paddle wheel is firmly clamped to the crank and cannot turn independently. Drill two ⅙-inch (4-mm) holes in the square washer in opposite corners and then insert a 1-inch (25-mm) screw in each hole. Weld the nut to the washer, and the end of the crankshaft to the nut.

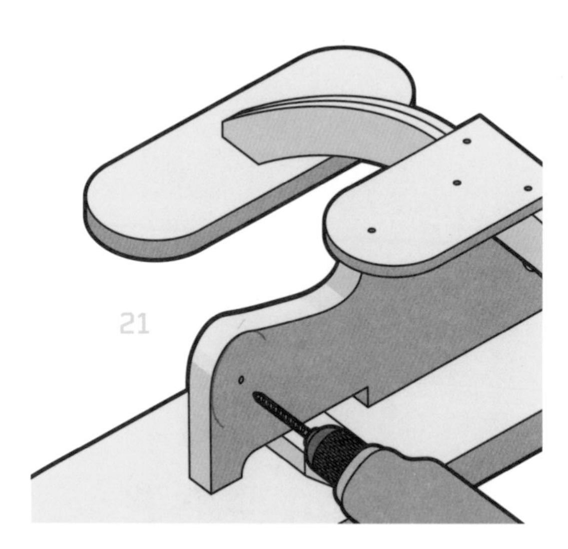

21 Drill a 5/16-inch (8-mm) hole through the seat support at the front, in the center point of the radius curve used earlier to mark out the seat.

22 From pine pieces E, make two levers with the top 6 inches (150mm) of each tapered with a block plane to make handles. To do this, secure the wood in a vise and, starting 6 inches (150mm) from the end, use a block plane, angled at 45 degrees, to make numerous passes across the wood, increasing the pressure as you plane over the end. Continue until the straight edge looks like one side of an octagon. Reposition the wood in the vise and repeat on the other corners so that the end finishes up octagonal.

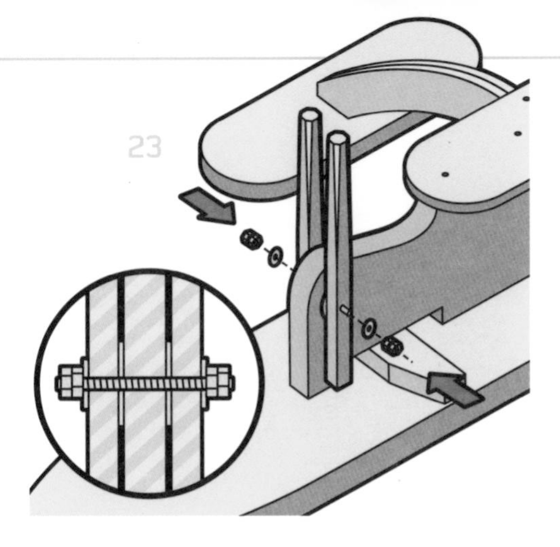

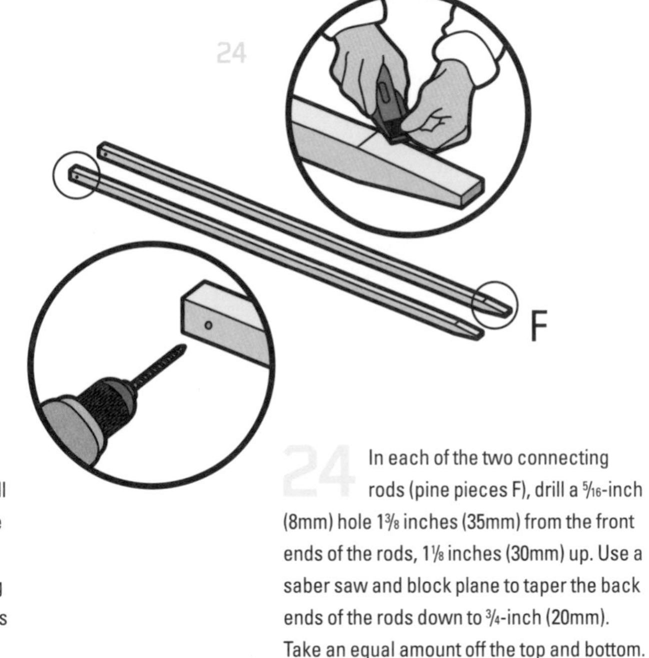

23 Using a flat bit, drill a countersink hole 1⅜ inches (35mm) in the bottom end of the hand-operated lever. This hole will accommodate the washer and head of a bolt, making a flush surface facing toward the seat support and hidden from view. Drill a ¼-inch (6mm) hole 1⅜ inches (35mm) from the bottom of each lever (aligning this with the countersunk hole) and another hole, the same size, 5 inches (125mm) above this first hole. Bolt the levers onto the seat support.

24 In each of the two connecting rods (pine pieces F), drill a 5⁄16-inch (8mm) hole 1⅜ inches (35mm) from the front ends of the rods, 1⅛ inches (30mm) up. Use a saber saw and block plane to taper the back ends of the rods down to ¾-inch (20mm). Take an equal amount off the top and bottom.

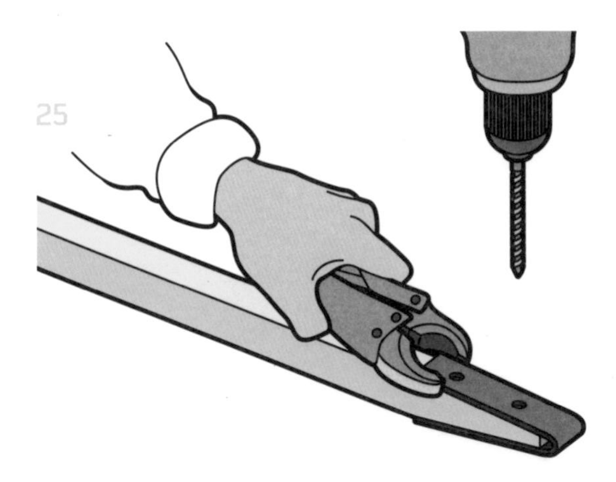

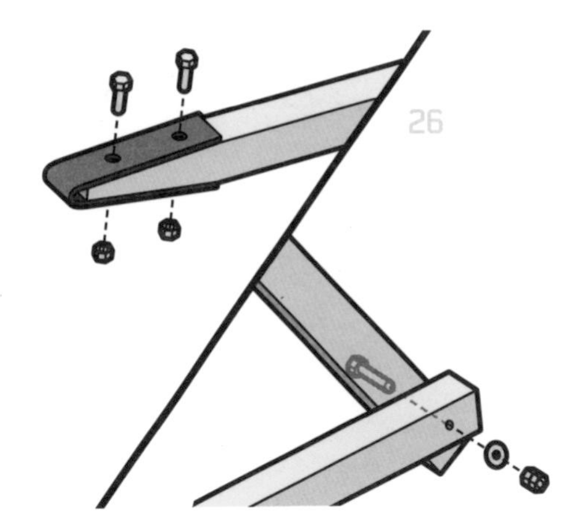

25 Using a vise and hammer, bend steel pieces M to make two loops to fit the tapered ends of the connecting rods. Drill ¼-inch (6mm) holes through the loops and rods at the same time, one at ½ inch (15mm) from the end of the loop (not the bent end, the cut end) and another 1½ inches (40mm) from the end (marking these holes with a center punch and hammer makes it easier and more accurate).

26 Fit the loops around the crankshaft and, using 2-inch (50-mm) bolts, and attach them to the ends of the connecting rods. Bolt the front of the connecting rods to the bottom of the handles using 3⅛-inch (80mm) bolts and nuts.

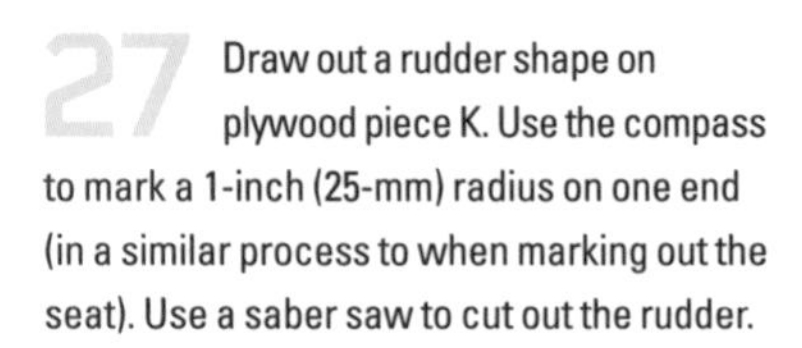

27 Draw out a rudder shape on plywood piece K. Use the compass to mark a 1-inch (25-mm) radius on one end (in a similar process to when marking out the seat). Use a saber saw to cut out the rudder.

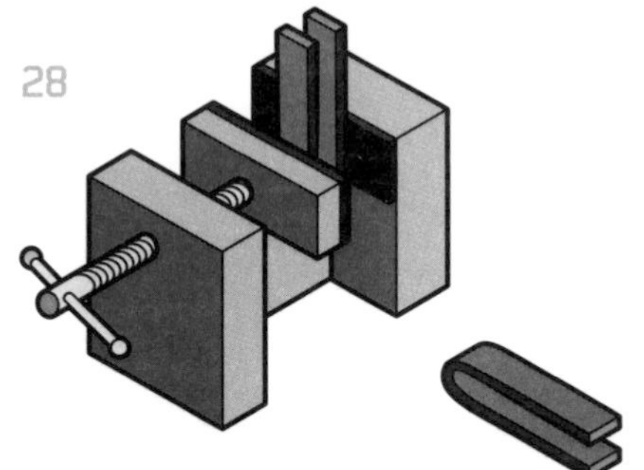

28 For the loop connecting the rudder, make two loops from steel pieces N, following the procedure outlined in step 25. Do the final bending, as shown, in a vise. Drill holes in the loops with a ¼-inch (6-mm) drill bit, ½ inch (12mm) and 2 inches (50mm) from the end.

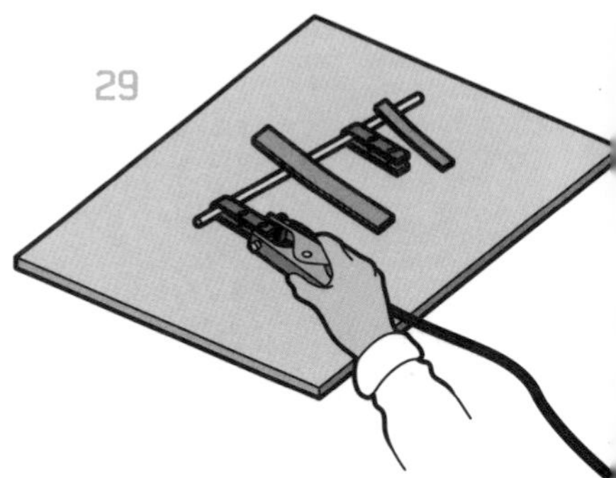

29 Set up for welding on a fire-resistant surface. Position the loops around the 14¾-inch (375-mm) length of threaded rod, one about 1 inch (25mm) from the end of the rod, the other about 7½ inches (195mm) from the same end. Use a file to remove any lacquer or galvanized finish from the metal where the welds will be made (the area where the loops touch the rod). Use cutoff metal as weights to hold the loops in the correct position. Clip the electrode to one of the loops.

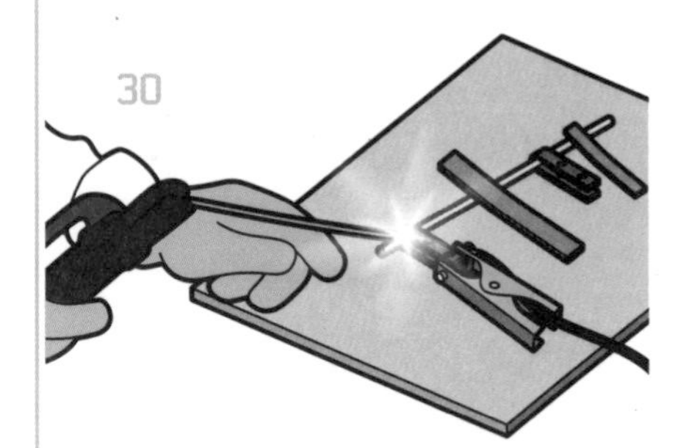

30 Tap the welding rod on the site to be welded to start the arc. Once the arc is struck, maintain close proximity as you move the rod over the surface to be welded. Feed the rod into the weld as you work (the rod is consumed). One rod is ample for this whole project.

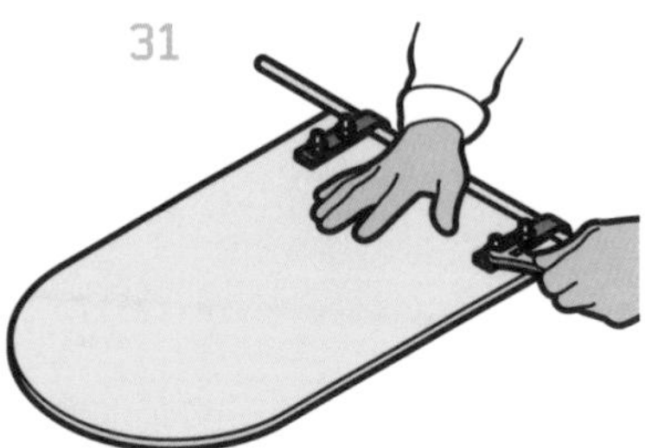

31 Bolt the loops and rod to the front end of the rudder, using 1⅜-inch- (35-mm) long, ¼-inch- (6-mm) diameter bolts.

32 Attach the rudder to the back of boat through a ⅜-inch (10-mm) hole, set 2 inches (50mm) from the back, centered on the hull. Position a large washer underneath the boat between the rudder and hull, and another above the hull. Fasten with two nuts.

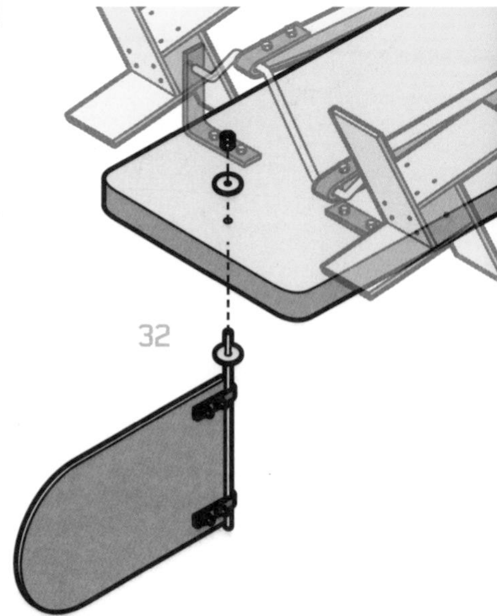

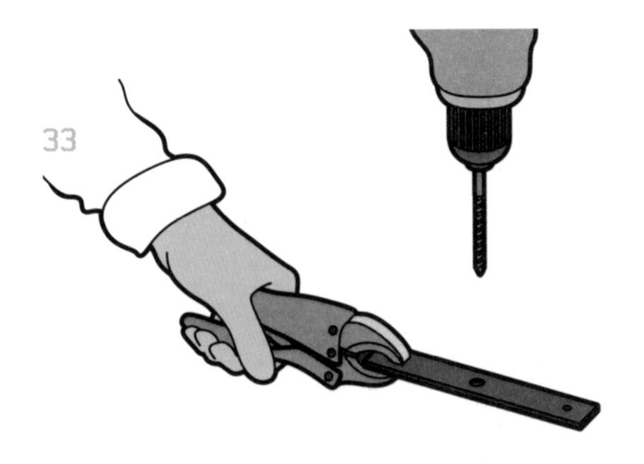

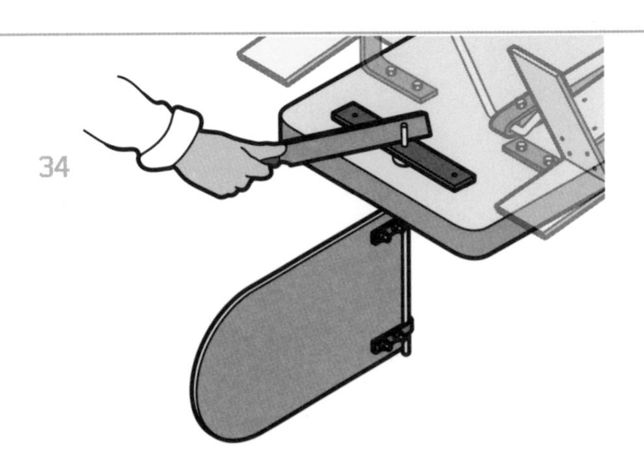

33 From steel piece O, make a lever (using a hacksaw and file) and drill a 3/8-inch (10-mm) hole in the center and 1/4-inch (6-mm) holes 1 inch (25mm) in from each end.

34 Place the lever over the rudder rod and file the areas of metal in preparation for welding the top surface of the lever to the rod, but leaving a 3/8 or 5/8-inch (10- or 15-mm) stub of threaded rod above the weld (the length of rod allows for this). Protect the surrounding area with noncombustible insulating material (for example, scrap metal) and weld around the joint. Let it cool. Put a dome nut on top of the rudder rod to give it a better appearance.

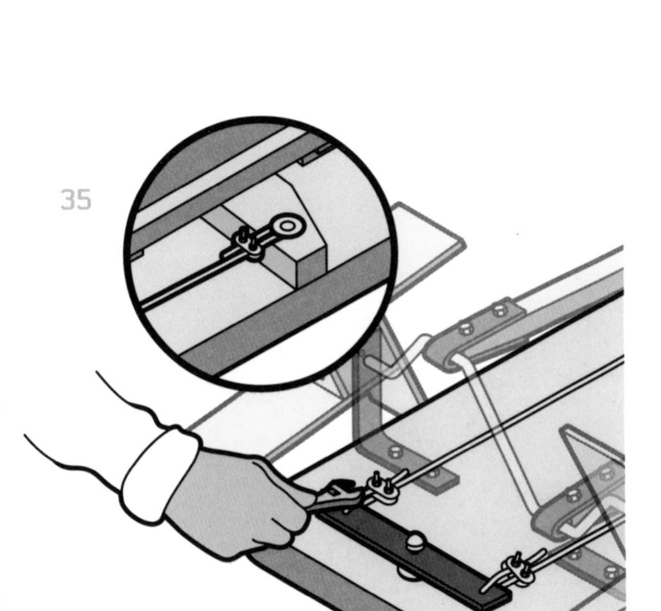

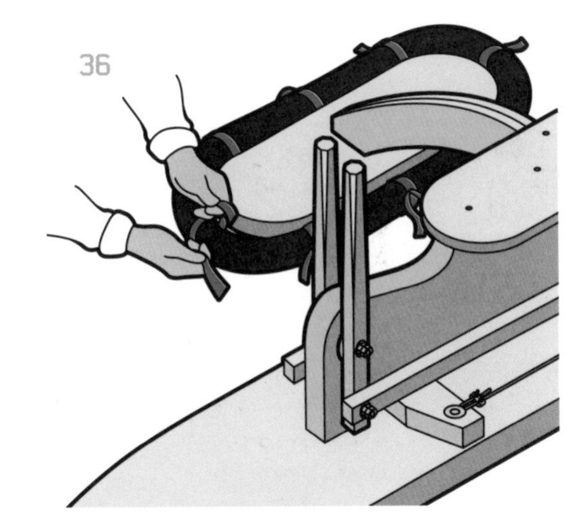

35 Attach one length of the wire rope to each end of the lever using wire rope clips, as shown. Attach the other ends of the wire ropes to the ends of the foot-operated lever at the front of the boat, pulling the rope tight and securing it with more clips, screws, and washers, as shown.

36 Apply two coats of marine varnish to the wood and the steel brackets to prevent corrosion. When the varnish is dry, screw lengths of webbing to edge of the floats (spaced evenly), inflate bicycle tire inner tubes, and tie the straps around the tubes. For extra buoyancy, cover the hull with a layer of foam insulation board in the same shape as the hull (use a handsaw to cut it). Glue it to the underside of the hull using silicon sealant. Likewise, cover the bottoms of the floats with a layer of foam.

The Paddle-Operated Punt

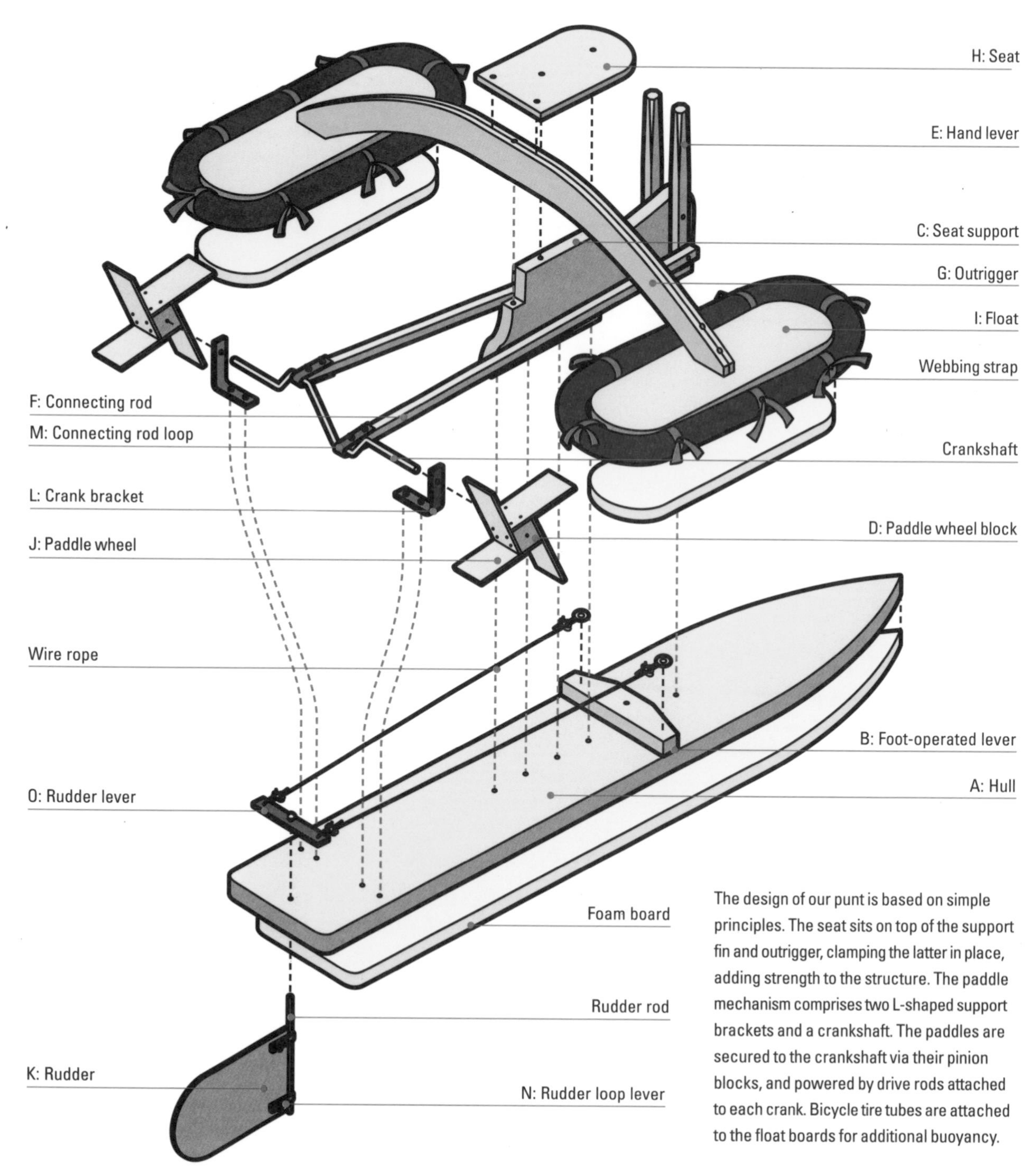

The design of our punt is based on simple principles. The seat sits on top of the support fin and outrigger, clamping the latter in place, adding strength to the structure. The paddle mechanism comprises two L-shaped support brackets and a crankshaft. The paddles are secured to the crankshaft via their pinion blocks, and powered by drive rods attached to each crank. Bicycle tire tubes are attached to the float boards for additional buoyancy.

How to Use It

To launch your punt, a dock or not-too-steep bank, some calm water, and a life jacket are all mandatory. Wear a wetsuit too, if you prefer to avoid getting too cold.

Getting onto the Punt

Tie a length of rope to a metal eye attached to the front of the hull, and tie the boat to the edge of the bank or dock to stop it from floating away. Lean over and hold onto the seat with one hand, then step onto the hull as near to the seat as you can. Next, hold the seat with both hands to balance your weight, before bringing your other leg onto the boat. The trick to keeping your balance when standing up on the boat is to always hold on with both hands. Get yourself seated comfortably, facing forward.

Peddle Power

Maneuver the hand-held levers back and forth in opposition; the alternating movements of the levers cause the crankshaft to rotate, which in turn rotates the paddles in either direction. It takes a little practice to start up the paddles going in the direction you want. Once they are set in motion, the paddles will continue rotating in that direction until you stop moving the levers. To get moving in the direction you would like, pump only one lever. Twist around and look at what the paddles are doing. When they are moving in the direction you want, face the front and start pumping the other lever.

Steering

To operate the rudder lever, use your feet. Depending on your height, you can use the sole of your foot or the back of your heel. When moving forward, pushing the left side forward will steer the boat left and pushing the right side forward will steer the boat right.

⚠ Safety

- Do not weld near combustible materials. Do not attempt welding before first receiving lessons from a competent professional. Wear only cotton clothing and cover all of your arms. Wear leather gauntlets and, preferably, a leather apron. Follow the welding machine manufacturer's instructions carefully. Never weld without wearing a welding mask (otherwise you will cause serious damage to your eyes). Do not allow children nearby while welding and do not allow anyone in the vicinity if they are not wearing a welding mask. After welding, do not touch the metal: it will be extremely hot and will burn you. Let it cool before continuing work.
- Wear goggles and a mask when using the saber saw and when drilling.
- Always file down metal after cutting with a hacksaw; otherwise the sharp ends are likely to cause cuts.
- Wear a mask when sanding.
- Wear gloves when bending metal.
- Wear a life jacket when using the boat and prepare for your feet and legs getting wet.
- Launch in calm waters with a friend on shore able to give assistance or call for help if you get into trouble.
- Consider what may happen if you or a child on the boat drift away from the shore, or the boat malfunctions and you are unable to operate it. For beginners and children, tie a long length of rope to the boat so that it can be pulled back to shore, if necessary.
- Keep an oar tied to the deck for emergencies and maneuvering through tight spots.
- If you are unfamiliar with boating, there will be dangers you are unaware of, so enlist the help of someone with experience.
- If you are unfamiliar with a stretch of water, ask the advice of locals before assuming that it is safe to use the water. Observe danger notices; tidal or shallow water, hidden rocks, mooring ropes, and mud banks can be dangerous or problematic.

Pneumatic Boat

Loaded with retro charm, this little boat has old-fashioned tugboat looks and a putt-putt chugging motion through water. It's also very simple to make, so if you have the urge to make something but only a couple of hours to make it in, this project is a good choice. When the balloon is inflated and fitted over the funnel, the air passes through a copper pipe which goes down through the keel and emerges below the boat. Crimping the pipe ensures that the exit point for the air is narrower than the entry point and gives the boat's passage through the water some force and speed. Ensure that you sand the funnel smooth, as any roughness here will puncture the balloon as it is fitted into position—and could bring your boat's maiden voyage to an abrupt end.

YOU WILL NEED

- Prepared knot-free pine, 18 × 6½ × ¾ inches (450 x 165 × 20mm)—this length allows for cutting waste
- Sheet of cardboard, 11 × 8½ inches (215 x 280mm)
- 6-inch (150-mm) length of 1-inch-(25-mm) diameter dowel (you can use an old wooden broom handle)—this length allows for cutting waste
- 12-inch (300-mm) length of ⅜-inch-(10-mm) diameter soft copper pipe—the type used for easy-to-bend micro plumbing
- Package of big, strong balloons

TOOLS

- Clamps
- Compass (geometrical)
- Cordless drill/driver with a selection of drill bits to fit, including a Forster bit—1 inch (25mm) in diameter
- Craft knife
- Emery paper
- Pencil, ruler, and square (for measuring and drawing lines)
- Pliers
- Saber saw with a good selection of "medium" blades to fit
- Sandpaper (fine grade) and sanding block
- Scissors
- Small file
- Straight handsaw
- Waterproof wood glue
- Workbench with vise

How to Make the Pneumatic Boat

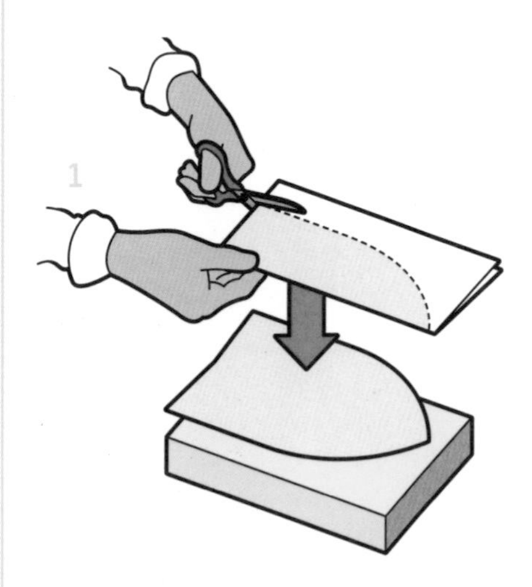

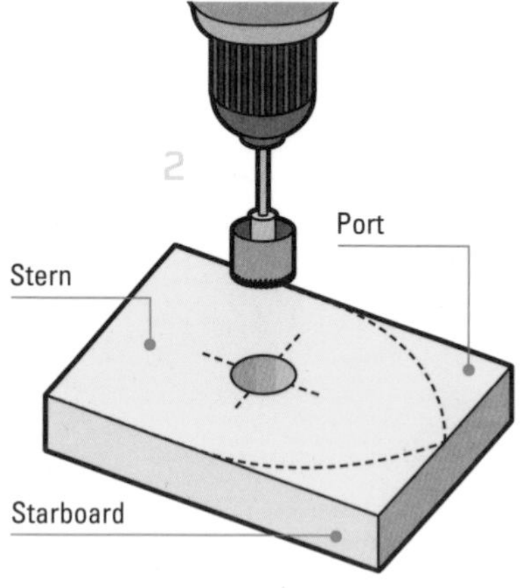

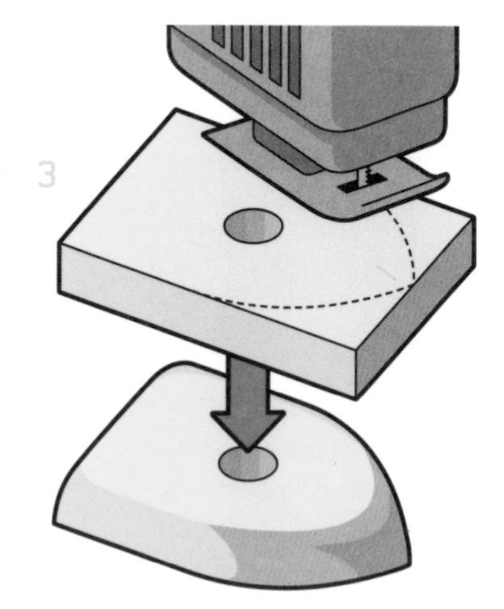

1 To make the hull, saw the pine to length so that you have a piece of wood 10½ inches (270mm) long and 6½ inches (165mm) wide. Use the template on page 155 to draw the outline of the hull on to the wood.

2 While you still have square sides to clamp, take the opportunity to drill your funnel hole. Use a 1-inch (25-mm) diameter Forstner bit to run the funnel hole through the thickness of the wood with its center 5 inches (125mm) from the stern and central to port and starboard.

3 Now use a saber saw to cut away the profile of the bow. Make sure that the line of cut is a little to the waste side of the draw line. Once you have your basic shape, you can use files and sanding blocks to shape the under-surface if you want your boat to have a more streamlined look.

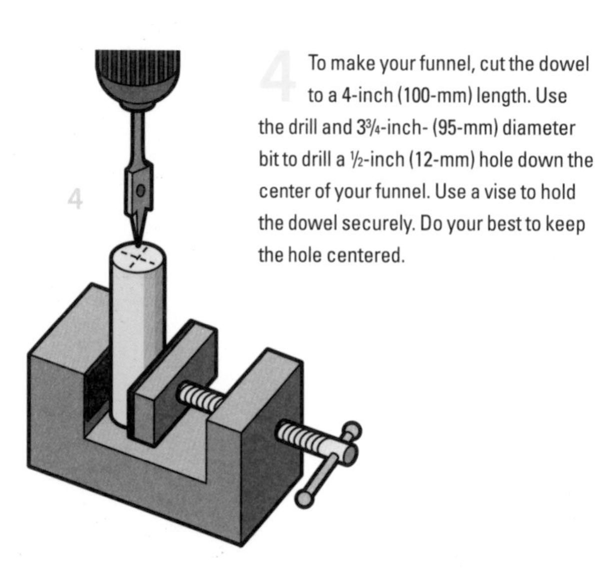

4 To make your funnel, cut the dowel to a 4-inch (100-mm) length. Use the drill and 3¾-inch- (95-mm) diameter bit to drill a ½-inch (12-mm) hole down the center of your funnel. Use a vise to hold the dowel securely. Do your best to keep the hole centered.

5 Use a knife and sandpaper to shape the top of the funnel. Round the edges, so that a balloon rim will slide over without to much difficulty. Then make a circumferential groove, about ½ inch (12mm) from the top, sufficient for the balloon rim to sit in without coming loose.

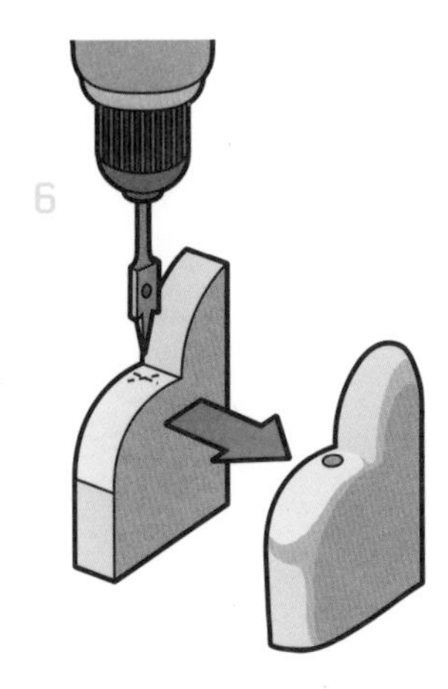

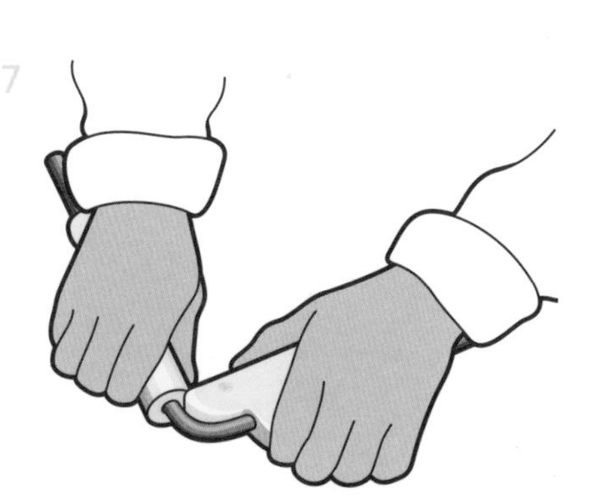

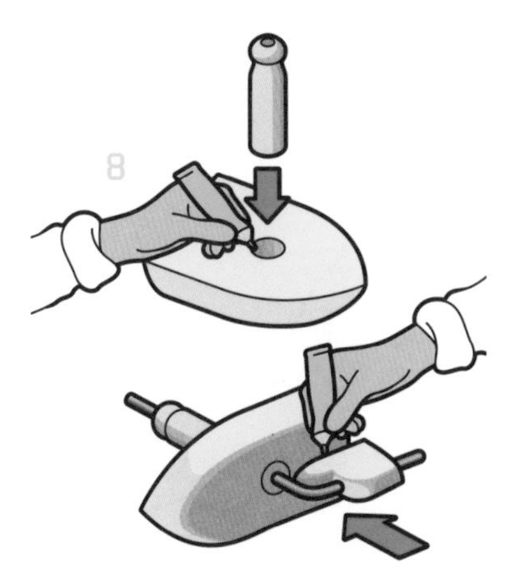

6 Use a pencil, ruler, and square to draw the shape of the keel block on the remaining piece of ¾-inch (20-mm) thick wood. It needs to be a stepped, fin shape, as illustrated; 4 inches (100mm) long, 2 inches (50mm) wide, with a 1-inch (25-mm) radius, quarter circle curve at one end. Saw the shape of the keel block out using a saber saw. Before shaping it, tip it on its back and drill a ½-inch (12-mm) hole through the keel block, in the same way that you drilled the funnel.

7 The final component is the copper propulsion pipe. Pass the pipe down through the funnel, starting at the top end, on through the curved end of the keel. Then, very carefully fold, ease, and hinge the two wooden parts together, bending the pipe over the keel. When you are happy with the shape, use pliers to snip the pipe off what will be the stern of the boat.

8 Assemble your pneumatic boat by first gluing the funnel into its recessed hole in the hull. Then glue the underside of the hull to the uppermost side of the keel, making sure that the pipe sits well along the funnel. Clamp and let the glue to dry.

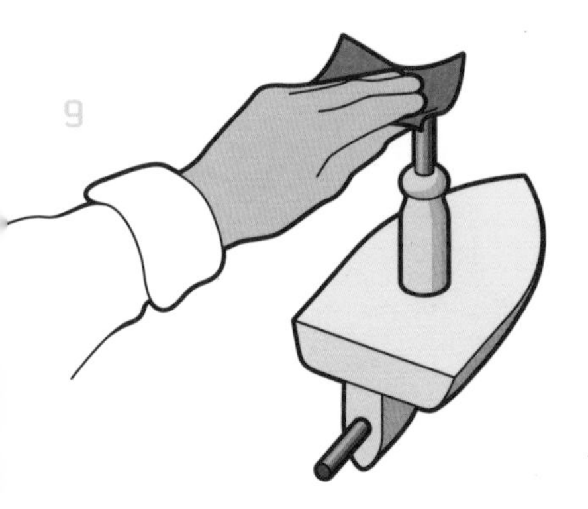

9 The pipe protruding from the funnel must be very smooth to ensure that it doesn't pierce the balloon, so smooth the ends with a metal file.

How to Use It

To test your boat, first inflate your balloon. While clamping the neck of the balloon with one hand, use the other to fit the rim over the funnel. Let go of the balloon and see how long it takes to deflate. You can adjust the opening of the tailpipe with pliers so the air is released at a slow, steady rate. Now you're ready to test-sail your boat on some water.

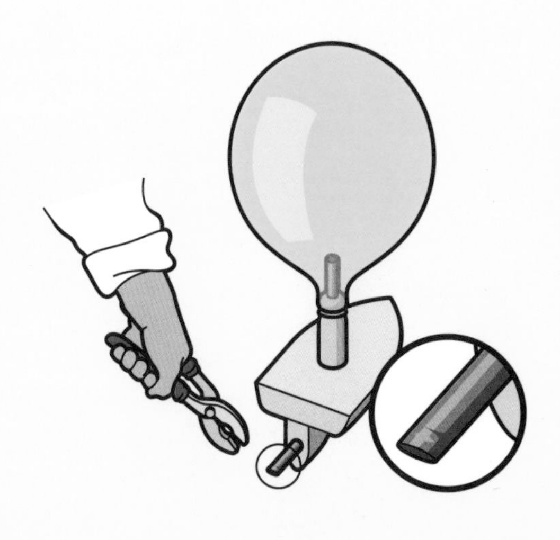

Hot Air Balloon

The most romantic project in the book, the hot air balloon would be a great way to mark a special day. It needs two or three people to launch it, too, so its maiden flight is worth making a fuss about. Keep a camera on hand to record it floating slowly up into the sky. From a short distance away, it will look like the full-blown, passenger-carrying version.

Large but quite delicate, with a tissue-paper body and a beverage-can "basket", the balloon needs only neat cutting and gluing rather than more complex skills. However, you'll need a fairly large workspace to lay the pieces flat, though, and you should launch your completed model with caution, because the air that propels it skyward is warmed by highly flammable denatured alcohol.

YOU WILL NEED

- Twenty-five sheets of white tissue paper, 25½ × 19¾ inches (650 × 500mm)
- Twenty-five sheets of red tissue paper, 25½ × 19¾ inches (650 × 500mm)
- Diluted white glue
- Cardboard boxes (for making the template and jig)
- Roll of duct tape
- Plastic sheet (thick garbage bags are suitable)
- Brads
- Bucket or large paint container (for sealing the balloon top opening)
- 60-inch (1,500-mm) length of ⅛-inch- (4-mm) diameter fiberglass rod (available from boating or building suppliers)
- 4-inch (100-mm) length of ¼-inch- (6-mm) diameter bore brass tube, (available from plumbing or building suppliers)
- Reel of strong thread
- Empty beverage can (a plain silver one will look best)
- Nail (for making holes in the can)
- 60 inches (1,525mm) of 24-AWG- (0.5-mm diameter) copper or galvanized steel wire
- Bottle of denatured alcohol
- Box of long matches
- Spool of 2½-lb (1.15-kg) test fishing line

TOOLS

- Adjustable work bench
- Awl
- Craft knife
- Hacksaw
- Hairdryer
- Heavy scissors
- Large flat and clean work space
- 1-inch (25-mm) Paint or glue brush
- Pen
- Pencil
- Pliers
- Protractor
- Sandpaper (medium-grade) and sanding block
- Steel ruler
- Tape measure

How to Make the Hot Air Balloon

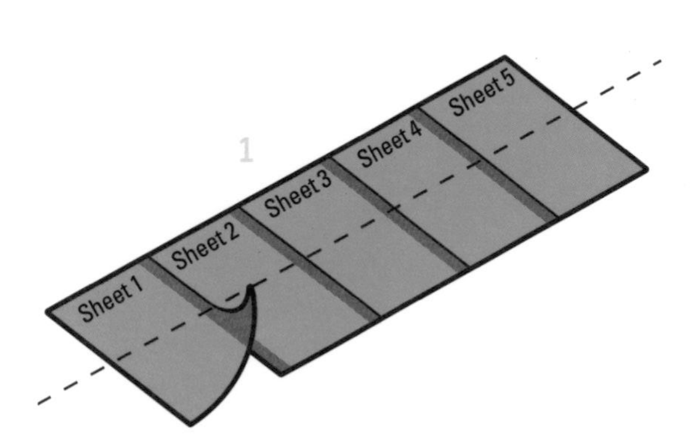

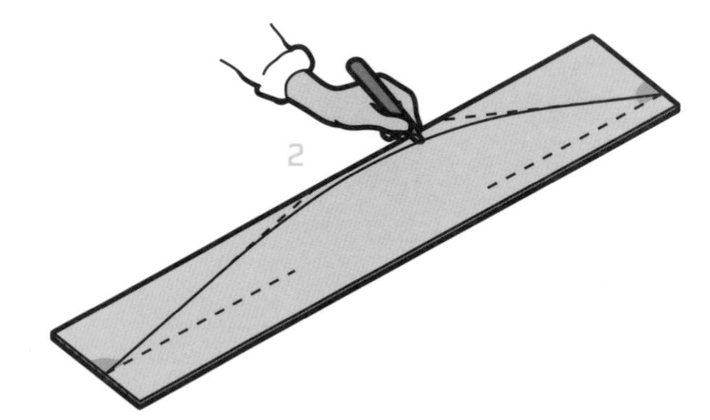

1 Separate each color tissue paper into five stacks of five sheets. Take the first stack, separate the top sheet, and spread glue along one of its edges. Place the next sheet on top of the first, edge-to-edge, leaving a 1 inch (25mm) overlap. Repeat until you have a strip of five sheets. Leaving this first strip to dry, repeat the process another nine times until you have five strips of red and five strips of white. When the glue is dry, fold the ten rectangles exactly in half along the midline.

2 Use a sheet of cardboard to make a template measuring roughly 90 inches (2,285mm) long and 13 inches (330mm) wide—the same width as the folded tissue sheets. Using the dimensions and angles shown on the template on page 155, draw lines on the cardboard at each end. Connect the lines together by using a bent steel ruler, on edge, to create a curve for your pencil to follow, or else draw a line by eye. Cut your template out, with scissors or a craft knife.

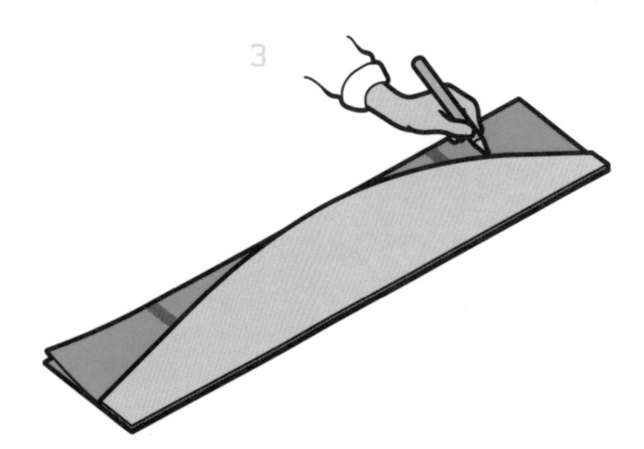

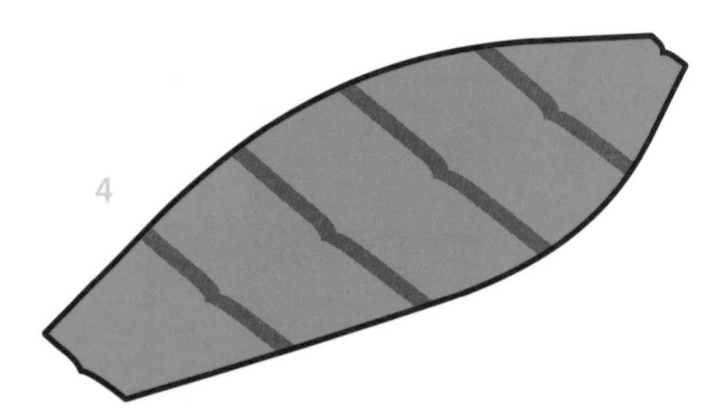

3 Using your cardboard template, you can now mark and cut the tissue panels so that they are all identical in shape and size. Use the template to guide a pencil over the tissue and then cut the panels out with heavy scissors, making sure that the fold is always on the opposite side to the cut.

4 The remaining pieces of tissue paper can then be unfolded to reveal petal shaped panels. The cardboard template can now be converted into a makeshift jig for assembling the ten flat panels into the three-dimensional envelope for the balloon.

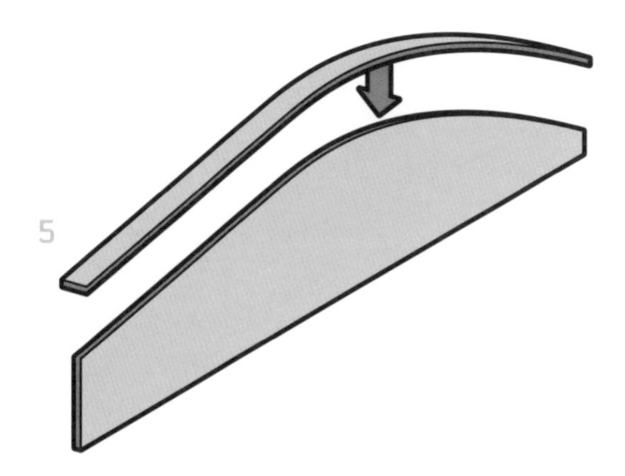

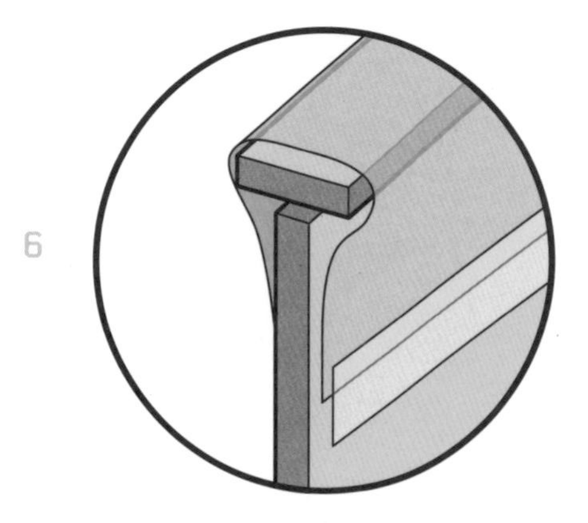

5 To make a jig from the template, use glue or duct tape to secure a 2-inch (50-mm) strip of cardboard perpendicular (90 degrees) to the template, along the entire curved edge of the cardboard template. This will act as a curved surface for gluing the tissue panels together.

6 Wrap a 6-inch (150-mm) plastic strip over this surface. Tape the plastic on one side and pull it taut before taping down on the other side—this will prevent the tissue from sticking to the jig. Now, set your jig in an upright position by clamping it in an adjustable work bench.

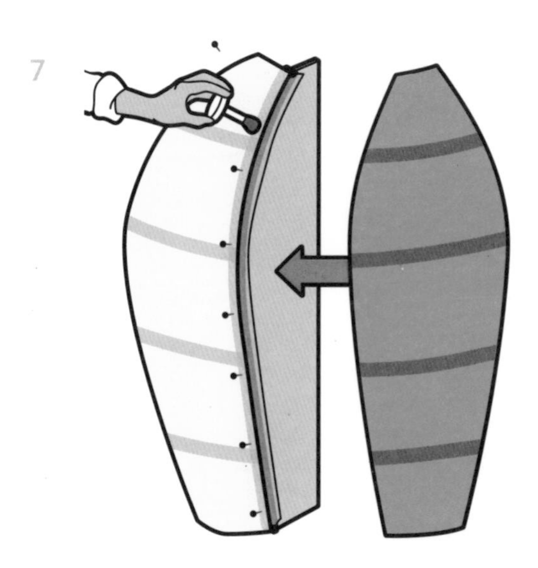

7 Pin the edge of a red tissue panel along the cardboard strip from top to bottom, and apply a layer of glue. Gently place the edge of a white panel over the glue and pat down with about 1 inch (25mm) overlap. Start at the top and work your way downward, removing and replacing the brads as you go along, until both panels are completely adhered together along the seam.

8 When the glue is dry, remove the brads and repeat the process, adding a third panel to the second, a fourth panel to the third and so on, until all ten panels are attached together. Glue panels of alternate colors together each time. To make the final join between the first and the tenth panels, ask a friend to hold the jig at one end, letting the balloon envelope surround the jig and meet up at the final seam. When the final seam is glued and dry, simply remove the jig through the aperture at the bottom.

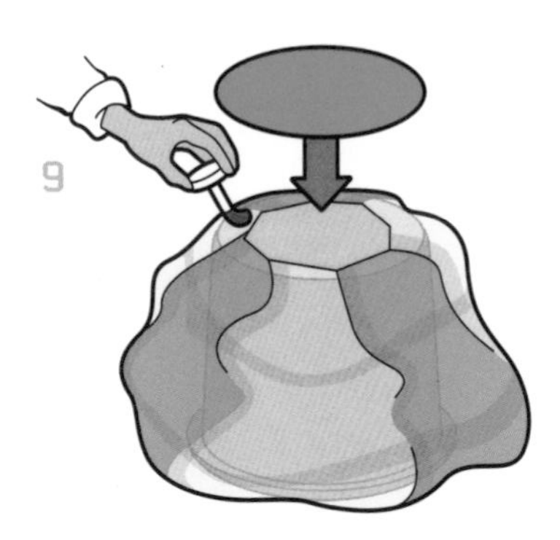

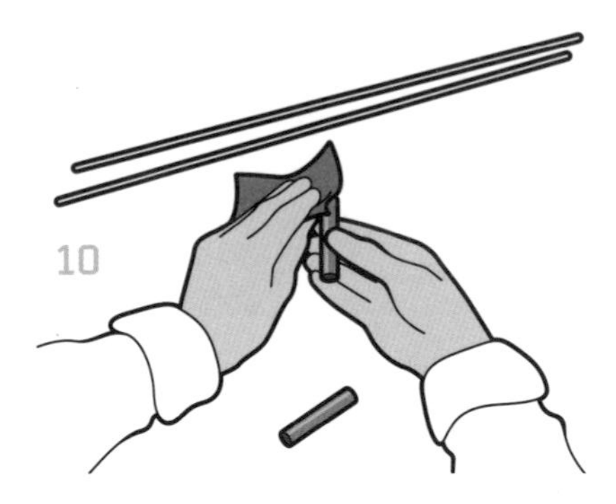

9 Sit the rim of the opening over the base of an upturned bucket, so that a flat surface supports the glued seam from beneath. Draw a disk of red tissue 12 inches (300mm) in diameter. Spread glue over the edges of the opening, lay the disk over it, smooth it down and leave it to dry. If the surface beneath is plastic the tissue shouldn't stick to it, but you can always pull some polythene over it to make certain.

10 Using a hacksaw, cut the fiberglass rod into two lengths, each 30 inches (760mm) long. Saw the brass tubing into two 2-inch (50-mm) lengths. Smooth the ends of the brass tubing with sandpaper. Gently bend the fiberglass around and use the tubes to connect them into a ring shape. This is the support ring which will reinforce the bottom opening of the balloon envelope, so that it holds itself open and can support the weight of the fire basket.

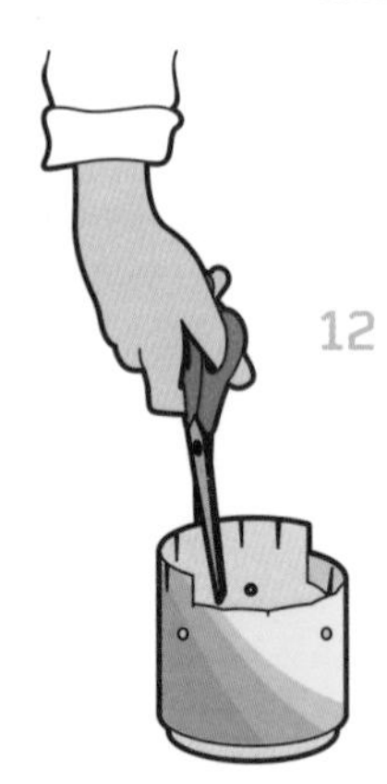

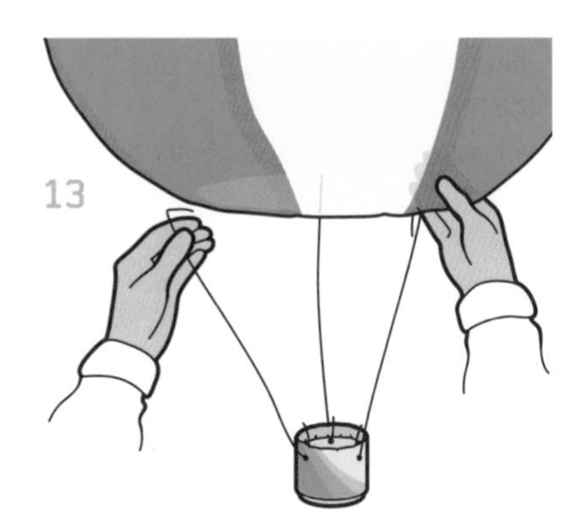

11 Position the hoop within the opening of the balloon envelope. Fold the tissue over it so that it encases the hoop, then snip the hem at 2-inch (50-mm) intervals, so that it forms a number of flaps. Hang the hoop on three strips of strong cotton thread and let the envelope sit evenly on the hoop before gluing the flaps in place. You may need to adjust the diameter of the hoop slightly to achieve a good fit. When the glue is dry, remove the threads and your balloon envelope will be complete.

12 Now make the burner. Cut the empty beverage can in half with a strong pair of scissors, being careful to avoid cutting yourself. Snip the sharp edge of the half-can at regular intervals and equal depth with the scissors and use pliers to bend the flaps over, crimping smooth so that there is no longer a sharp edge. Using a protractor, mark the top of the can at three even points, at 120-degree intervals. Mark the points with a pen, then use a nail to make a hole at each point about ¼ inch (6mm) down from the edge.

13 Cut three 18-inch (460-mm) lengths of 24-AWG- (0.5-mm diameter) copper wire. Attach one end of each through one of the holes in the can, by passing it through the hole and crimping it tightly over. Fasten the free ends of the copper wires to the support hoop in the mouth of the balloon, by pushing the wire gently through the tissue and folding it around the fiberglass rod. Make sure the three wires are spaced at even thirds around the support hoop (approximately three-and-a-third ballloon panels apart).

How to Use It

For your balloon's maiden—and potentially only—flight, you need a very still, cold day, a wide open space (make sure there are no buildings, trees or overhanging wires nearby), and the help of a few friends for your launch. The latter shouldn't be a challenge; our launch attracted a crowd of 20, even without prior publicity.

Inflating the Balloon

Place the can burner on the ground ensuring that the hooks on the copper wires are crimped tight at both ends. Arrange two or three people around the sides of the balloon, to hold it out and upright. Carefully fill the aluminum can burner with denatured alcohol to a depth of 1½ inches (40mm). Hold the balloon directly above the burner, so that when lit, the flame goes up inside and doesn't catch the tissue paper. Take a long match or taper and light the denatured alcohol, while your helpers hold the balloon steady. After three or four minutes the balloon will start to fill out and gently rise. As it does so, hold the support ring. When the balloon is inflated and pulling against your hand, that is the sign the balloon is beginning to rise under its own convection, and is ready for its flight. Give the signal to let it go free. Instead of using alcohol to heat up the air inside the balloon, you can also use a blow dryer or blow torch (if using the latter, make sure the flame doesn't catch the tissue paper).

Extending the Balloon's Shelf Life

There is nothing more magical than seeing the balloon ascend and disappear off into the sky, left to its own devices. If you would like to witness the spectacle again and again, you can tether your balloon to the ground with a length of fishing line attached to a heavy object. You will need to tie the other end of the fishing line to the reinforcement hoop, so that it isn't melted by the flame. The balloon will descend once the denatured alcohol has burned off. You can fly the balloon inside, but make sure that the tether is the correct length and be careful that the can burner does not spill its contents.

Proceed with caution when using a blow torch.

⚠ Safety

- Even if you follow these directions, there is a minimal risk of the tissue catching fire. Should that happen, you and your helpers should walk away immediately to a safe distance, and leave the balloon to burn out.
- Don't launch your balloon in any areas where there is a fire danger. Keep away from trees, especially when the wind is gusting. Keep an eye on the balloon throughout its flight, as a basic safety precaution.

Free-Flight Airplane

There's something very satisfying about working in balsa wood: it offers all the joys of carpentry without any of the hard labor involved. The propulsion mechanism comes from a model store, leaving you with the work of cutting the pieces, conducting trial flights and amending elements to enhance its performance. Make several models, and you'll find that each one, however identical it looks to the others, will fly slightly differently. The glider's structure is simple enough for children to help with, or even make entirely themselves. If your family has a competitive streak, you can organize a contest for the best construction, the best physique, and, of course, the glider that flies the longest distance.

YOU WILL NEED

Balsa stick to make up:

- A: piece, $14\frac{1}{2} \times \frac{3}{8} \times \frac{1}{8}$ inches (370 × 10 × 3mm)—fuselage

Balsa wood sheet to make up:

- B: piece, $8\frac{3}{4} \times 3\frac{1}{2} \times \frac{1}{8}$ inches (220 × 90 × 3mm)—left wing
- C: piece, $8\frac{3}{4} \times 3\frac{1}{2} \times \frac{1}{8}$ inches (220 × 90 × 3mm)—right wing
- D: piece, $8 \times 3 \times \frac{1}{16}$ inches (200 × 75 × 2mm)—tailplane
- E: piece, $2\frac{5}{8} \times 2\frac{5}{8} \times \frac{1}{16}$ inches (65 × 65 × 2mm)—fin

- Balsa cement
- 9 inches (230mm) of 20-AWG- (0.812mm diameter) piano wire
- $5\frac{1}{2}$-inch (140-mm) plastic propeller with shaft and eye, and nose bearing with undercarriage slots, for fitting onto $\frac{3}{8} \times \frac{1}{8}$ inch (10mm × 3mm) fuselage—this is most common type sold in model stores
- Two 1-inch- (25-mm) diameter foam wheels
- Rubber band, $15\frac{3}{4}$-inch (400-mm) circumference
- Straight brad
- Plasticine (modeling clay)

TOOLS

- Adhesive tape
- Craft knife
- Long-nose pliers and pliers
- Metal ruler
- Pencil
- Razor saw
- Scissors
- Two sheets of thin cardboard or paper, $8\frac{1}{2} \times 11$ inches (215 × 280mm)—for templates

How to Make the Free-Flight Airplane

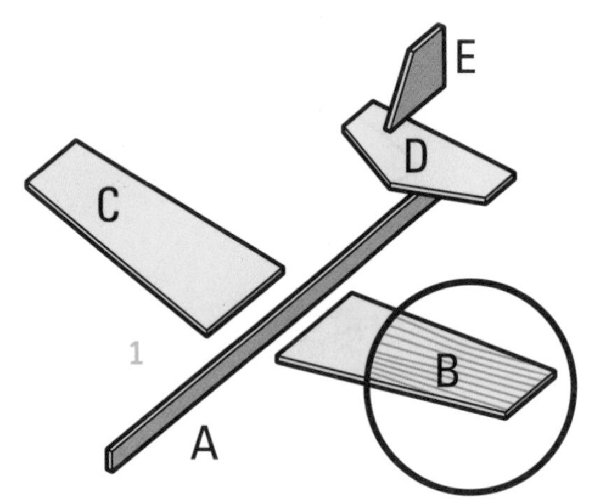

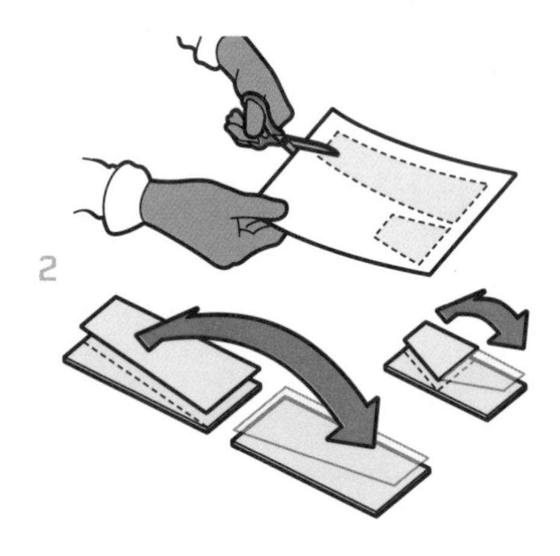

1 This airplane has only five balsa parts comprising its fuselage, wings, tailplane, and fin. Use the razor saw to cut fuselage piece A from the balsa stick, and a craft knife to cut the wings (pieces B and C), tailplane (piece D), and fin (piece E) from the balsa sheets, to the basic dimensions listed on page 68 (You Will Need box). Make sure the grain of the balsa wood follows the longest dimension of each component as much as possible.

2 To make certain that the wings and the tailplane are symmetrical, use cardboard or paper templates. The wings taper to 2½ inches (65mm) at the tips; the tailplane tapers to 1½ inches (40mm) at the tips, and the fin tapers to just under 1½ inches (40mm) at the top. For the wings, the same template can be used in reverse to mark out each corresponding wing.

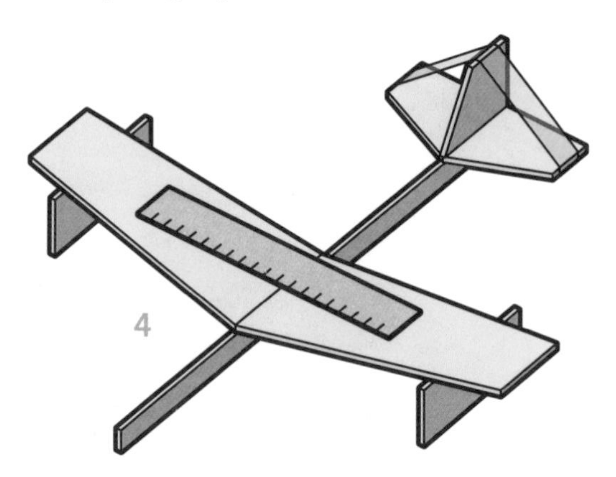

3 Now that you have the wings and tailplane cut to shape, you need to glue the wings together as one with balsa cement. Glue the wings with a slight angle to each other—about 155 degrees will do. Prop one of the wings up at 25 degrees while the balsa cement is curing.

4 Now you are ready to assemble your glider. Glue the tailplane onto the fuselage, sitting it flush with the end edge. Next, glue the fin onto the tailplane, again sitting it flush with the end edge. Stick the wings to the fuselage 3½ inches (90mm) from the front, using props of balsa to support the wing tips, and weigh down the wings using a metal ruler. Use adhesive tape to hold the parts in place while the joints bond properly. Only apply balsa cement where it is needed.

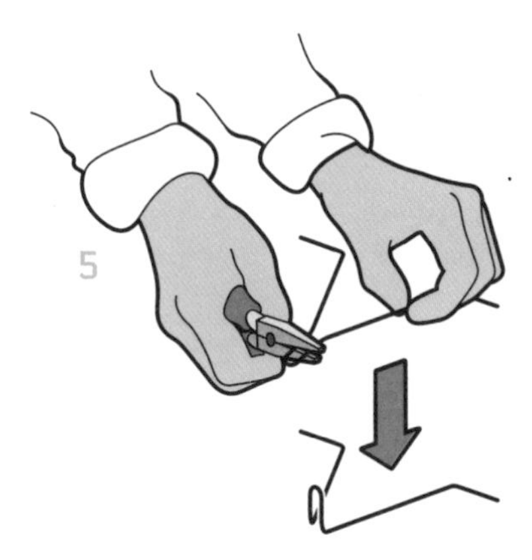

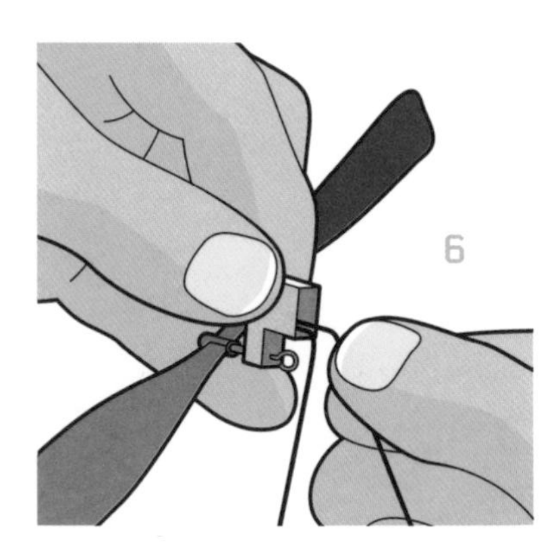

5 To make the undercarriage, use both pairs of pliers to bend the piano wire into a triangle shape, with the ends spaced 3 inches (80mm) apart. Bend the ends upward by about 90 degrees—the wheels will be attached to the ends later.

6 Turn the top ⅜ inch (10mm) of the undercarriage over by 120 degrees. Now you can link this turned-over wire to the propulsion mechanism, which is bought from a model store. Push the wire into the slot in the plastic nose bearing.

7 Push the plastic nose bearing onto the front of the fuselage piece. Attach the wheels to the turned-up wire ends of the undercarrriage, so that they are trapped in a vertical position (they can spin freely but cannot come off).

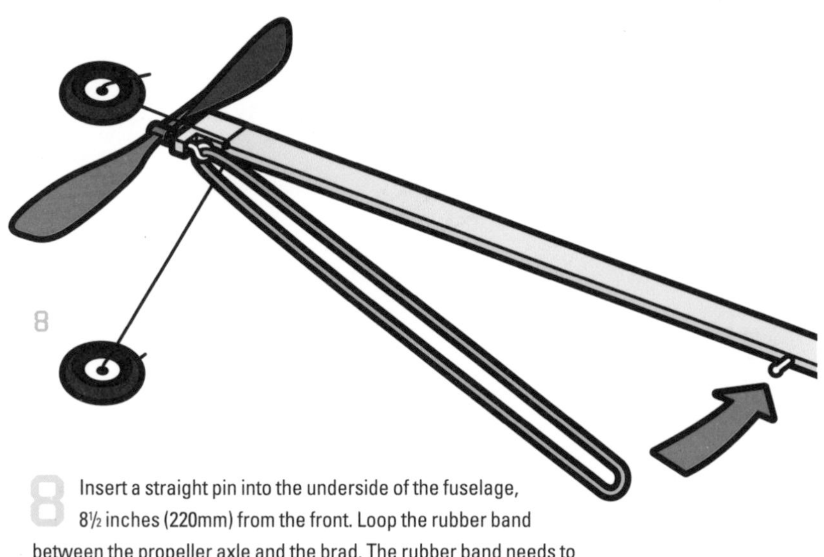

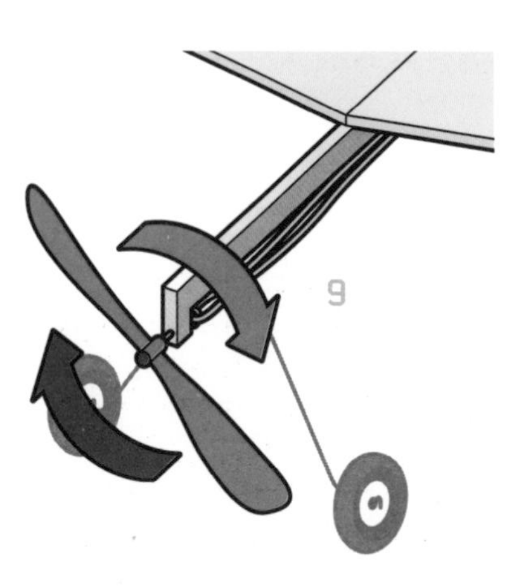

8 Insert a straight pin into the underside of the fuselage, 8½ inches (220mm) from the front. Loop the rubber band between the propeller axle and the brad. The rubber band needs to be slightly taut and remain in position.

9 To prepare your airplane for its maiden flight, wind the propeller in a clockwise direction (looking at the plane) roughly a hundred times, until the rubber band has twisted into knots along its entire length. Anything less that this and the airplane will be underpowered. Anything more and there is a risk of the rubber band breaking.

How to Use It

The rules for launching your airplane are not set in stone, so find a technique that suits you best. The main factor controlling the launch phase is the strength of your arm, so don't expect miracle launches as soon as you begin.

Launch the Airplane

Hold your airplane below the wings and lift it to shoulder height while holding the propeller with the other hand. Tilt the nose slightly upward, and the right wing slightly lower (assuming a right-handed thrower). Release the airplane with a steady forward motion after letting go of the propeller a split second beforehand. Release it into the breeze, if there is one, and make sure that there is nothing for it to collide with to avoid damaging it.

Setting off the airplane.

Tips

You can make adjustments to the aerodynamics of your airplane to address any problems with stalling, pitch, roll, and yaw. Steam from a kettle or saucepan can be applied to the balsa wood to soften it so that it can be bent slightly. The desired flight is a steady and straight climb, followed by a controlled glide.

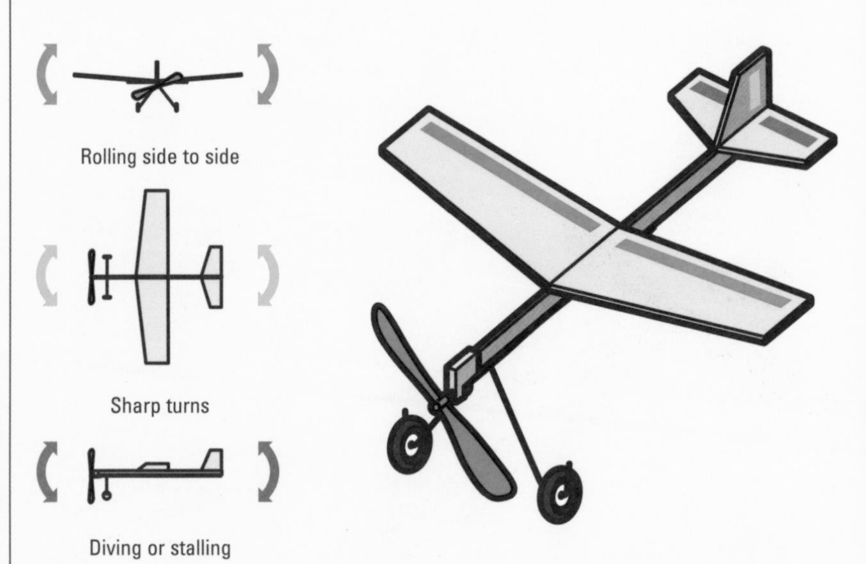

Once it is airborne, watch the airplane's attitude. While it achieves flight under its own propulsion, it is an airplane. When the propeller stops turning, it becomes a glider. You can adjust your wings, tailplane, and fin to make the most of your airplane. To stop it from rolling from side to side, glue a 1-inch (25-mm) long cardboard tab to the top surface of the right wing, centered on the trailing edge. If it dives slightly, put a plasticine (clay) weight on the back of the plane. To prevent sharp turns, reduce or remove the nose weight, and adjust the fin to bend more to the left or right.

MACHINES THAT WALK

Soapbox Go-Cart

Although it's scaled strictly for kids, the traditional action and box-on-wheels appearance of this go-cart is bound to create a few nostalgic sighs among adults. It calls for basic woodworking skills, but if you can cut a length of wood, you'll be able to make the whole cart without much trouble. Steering is operated by a simple bar-and-rope arrangement and the only hint of modernity in the whole design is four new wheels, which help to give the passenger a smoother ride than the more rough-and-ready, patched-up versions your grandparents might remember. Don't hesitate to recycle if you have any decent scrap wood or second-hand wheels that you can reuse. Old-style apple boxes, if you can find any, make great box seats, too.

YOU WILL NEED

Pine to make up:

- A: piece, 33½ × 2¾ × 1¼ inches (850 × 70 × 30mm)—centerpiece
- B: piece, 17¾ × 2¾ × 1¼ inches (450 × 70 × 30mm)—crosspiece
- C: piece, 17¾ × 2¾ × 1¼ inches (450 × 70 × 30mm)—crosspiece

- 2½-inch (65-mm) bolt, 5⁄16 inch (8mm) in diameter, with one nut and two washers (for securing the front crosspiece)
- 2-inch (50-mm) bolt, 5⁄16 inch (8mm) in diameter, with one nut and two washers (for securing the back crosspiece)
- Wooden crate, about 13¾ × 13¾ inches (350 × 350mm)
- Screws and/or brads to reinforce the crate
- Four 1¾-inch (45-mm) brads
- Three 3-inch (80-mm) carriage bolts, 5⁄16 inch (8mm) in diameter, with matching nuts and washers (for attaching the box)
- Two 40-inch (1,000-mm) lengths of ⅜-inch- (10-mm) diameter threaded rod
- Eight ⅜-inch (10-mm) washers
- Four 13-inch- (330-mm) diameter baby carriage wheels (or a similar substitute)
- Four ⅜-inch (10-mm) nuts
- Four ⅜-inch (10-mm) cap nuts
- Four 1 × 1 inch (25 × 25mm) corner braces
- Eight 1-inch (25-mm) screws (to attach corner braces)
- 4½-feet- (1,400-mm) length of cord or rope (for steering)

TOOLS

- Adjustable wrenches
- Cordless drill/driver with a good selection of drill bits to fit
- File
- Hacksaw
- Pencil
- Pliers
- Saber saw (or handsaw)
- Sandpaper and sanding block
- Scissors
- Small hammer
- Socket set
- Tape measure
- Try square
- White glue

How to Make the Soapbox Go-Cart

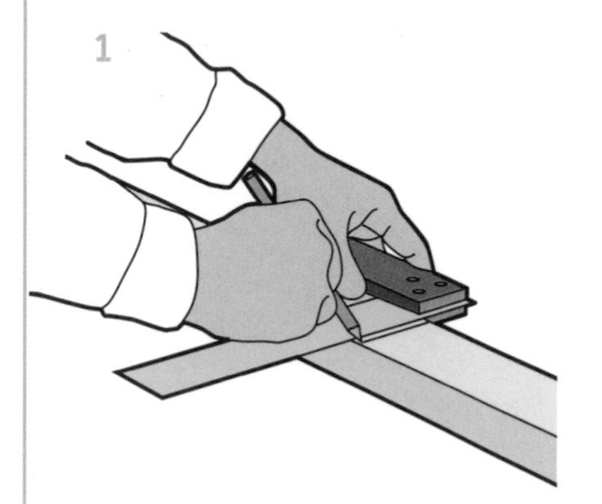

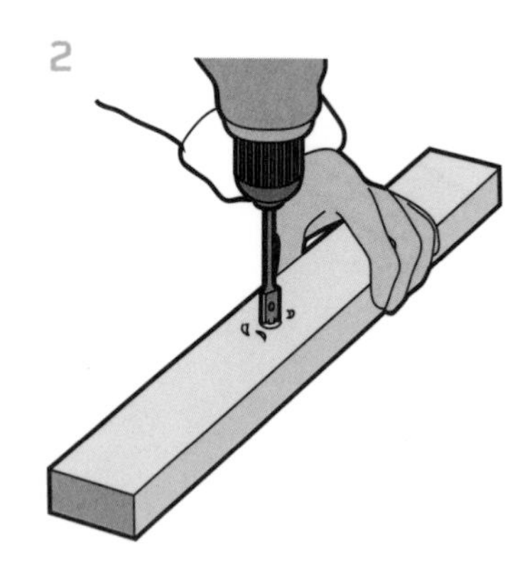

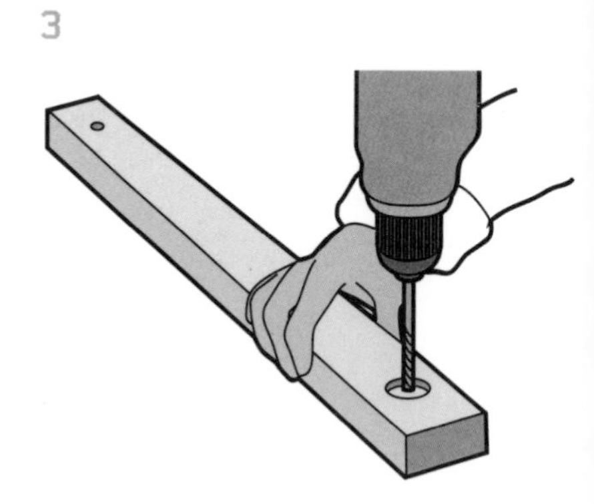

1 Mark the lengths of the centerpiece (piece A) and two crosspieces (pieces B and C) using a tape measure, pencil, and try square. Cut to length with a saber saw or handsaw. Sand the edges smooth with sandpaper.

2 Drill countersink holes, 3/8-inch (10mm) deep and 3/4 inches (22mm) wide in the center of both crosspieces. These holes will face downward and are recesses to accommodate a washer and a nut. In the center of the counterbore holes, drill 5/16-inch (8-mm) holes through the wood.

3 Drill a central countersink hole 3/8-inch (10mm) deep and 3/4-inch (22mm) wide in what will be the centerpiece at the back end of the cart, 1 3/8 inches (35mm) in from the end. In the center of that countersink hole, drill another 5/16-inch (8-mm) hole right through the wood, so that the head of a bolt will sit flush with the surface of the wood. Drill a single central 5/16-inch (8-mm) hole in the front end of the centerpiece 1 3/8 inches (35mm) in from the end.

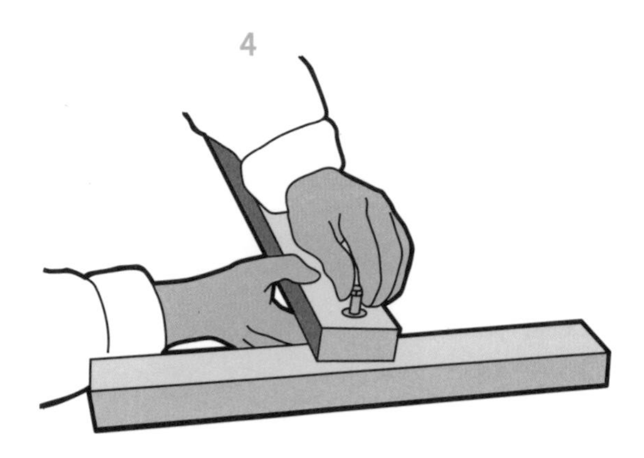

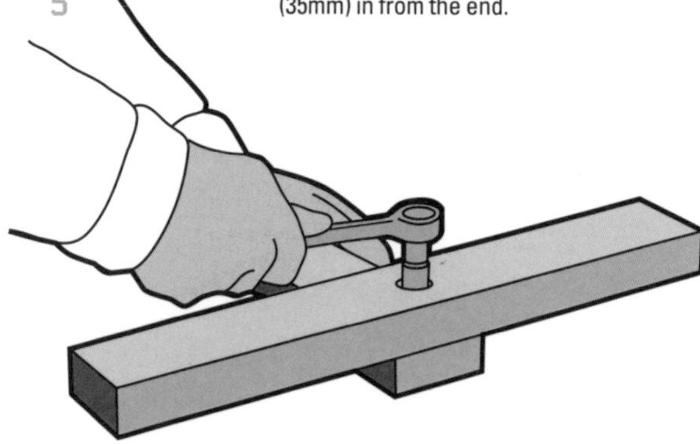

4 In the center of each crosspiece drill a countersink hole, 3/8-inch (10mm) deep, 1 inch (25mm) wide, followed by 5/16-inch (8-mm) holes all the way through the wood (the countersink holes are needed to accommodate the nuts—the nuts need to be flush with the surface of the wood so that the axle rods can pass over them).

5 Bolt the centerpiece to the crosspieces using washers under the heads and nuts (no need for a washer between the wood). Use a wrench to tighten the nut on the front crosspiece.

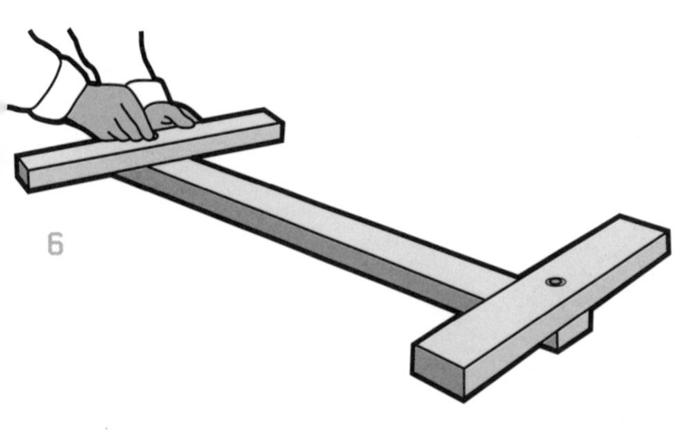

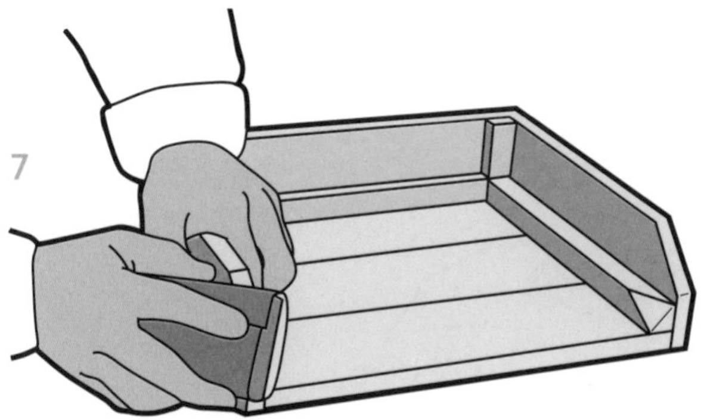

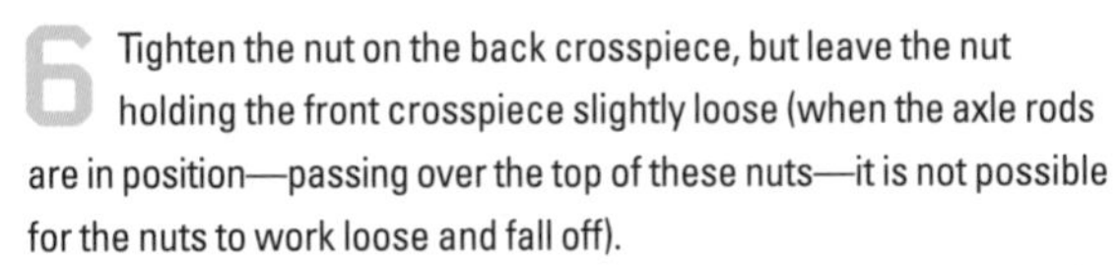

6 Tighten the nut on the back crosspiece, but leave the nut holding the front crosspiece slightly loose (when the axle rods are in position—passing over the top of these nuts—it is not possible for the nuts to work loose and fall off).

7 Prepare the wooden crate by removing one side and reducing the height if you want (sawing off, as necessary). Usually the crate will need reinforcing with additional brads or screws and possibly some cutoffs glued and nailed, especially if the sides appear to be fragile. Cut off the corners at the front if you like, check there are no nails protruding, and sand all over to round the edges and remove splinters.

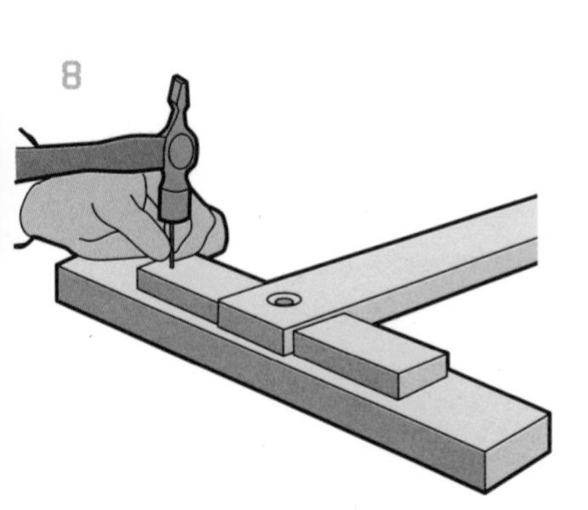

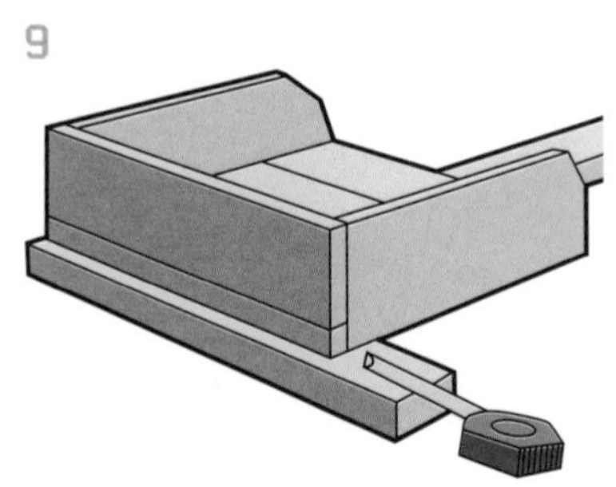

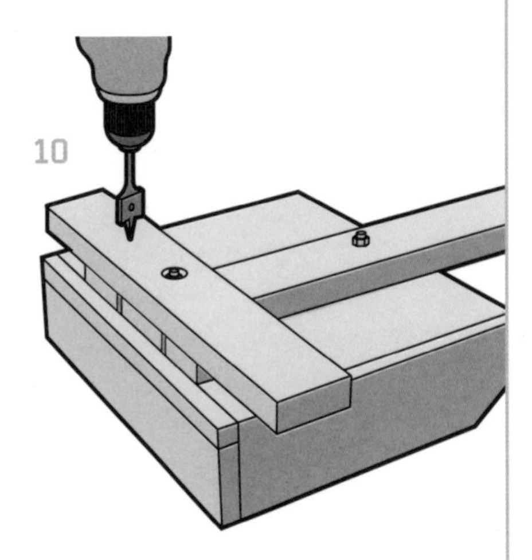

8 Glue and nail cutoffs of wood to the back crosspiece either side of the centerpiece using the small hammer. These need to be of a suitable thickness to support the box at the same height as the centerpiece.

9 Place the box centered on the centerpiece at the back of the cart (measure each side to check).

10 Drill $\frac{5}{16}$-inch (8-mm) holes for securing the box to the centerpiece and crosspiece. The nuts holding the crosspiece need to be flush (under the axle rod) so drill countersink holes first, $\frac{3}{8}$-inch (10mm) deep and $\frac{3}{4}$-inch (22mm) in diameter. Attach the box to the centerpiece and crosspiece using carriage bolts. Tighten the nuts.

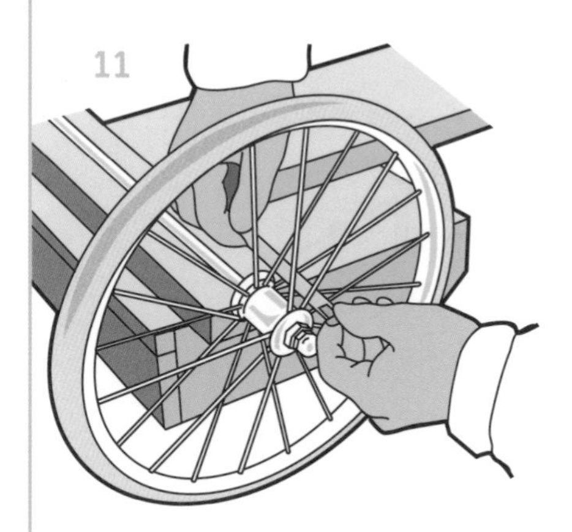

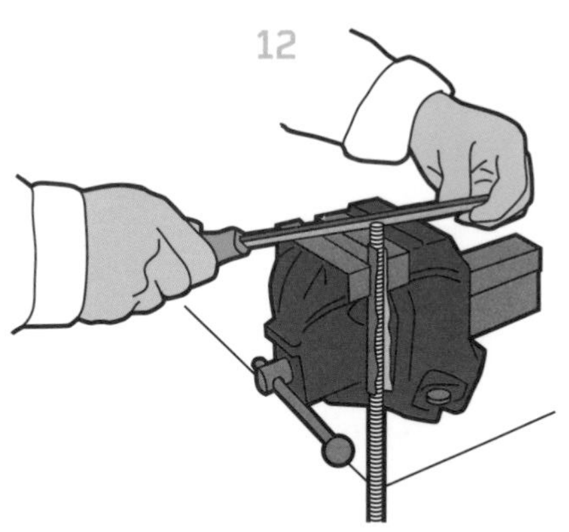

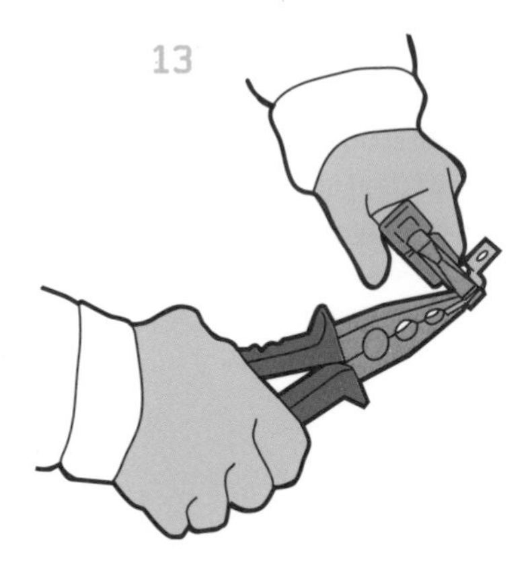

11 To work out the length of the axles, lay one of the lengths of threaded rod on a crosspiece, thread a large washer onto the rod followed by a wheel and another large washer, and screw on a nut and a cap nut. Fully tighten the cap nut; then measure the length of rod extending beyond the crosspiece.

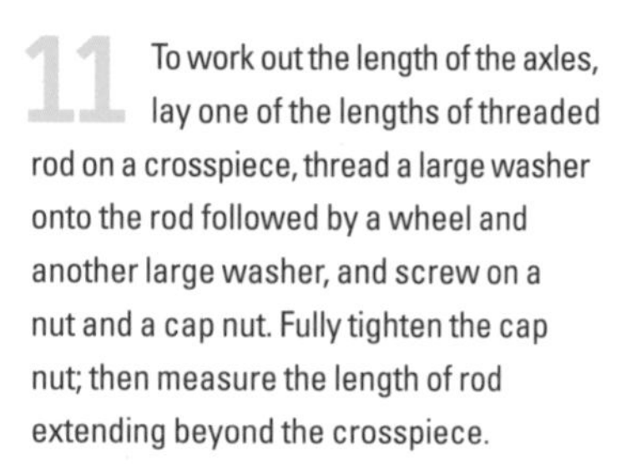

12 Allow for the same length on the other side, and then cut off any excess using a hacksaw. Use a piece of sandpaper to protect the threaded rod. File the ends smooth.

13 Use both pairs of pliers to reshape the corner braces to make the axle brackets.

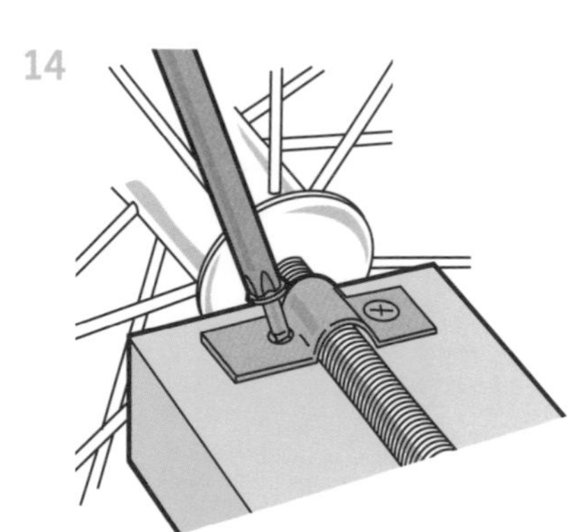

14 Fit the wheels to the axle rods using the washers and nuts as previously. Lay the axle rods along the center of the crosspiece and fit the brackets over the rod near the ends. Screw the brackets down.

15 For the steering cord, drill a small $\frac{5}{16}$-inch (8-mm) hole in each end of the front crosspiece, 1 inch (25mm) in from the ends and from the front edge. Thread the steering cord down through the holes in the front crosspiece and tie knots in the ends so that it cannot pull through.

⚠ Safety

- It is advisable to wear a bicycle helmet while using the go-cart.
- Do not use the cart on public roads or footpaths, where you run the risk of running into another person.
- It is very important to avoid steep hills; if you don't, you risk losing control and/or not being able to stop safely.
- Be careful when braking with your feet. Apply gentle pressure and avoid getting your foot twisted or trapped.

The Soapbox Go-cart

The simplicity of our go-cart can be clearly appreciated, especially the steering; the front crosspiece can be rotated left or right, via either the driver's feet, or the steering rope.

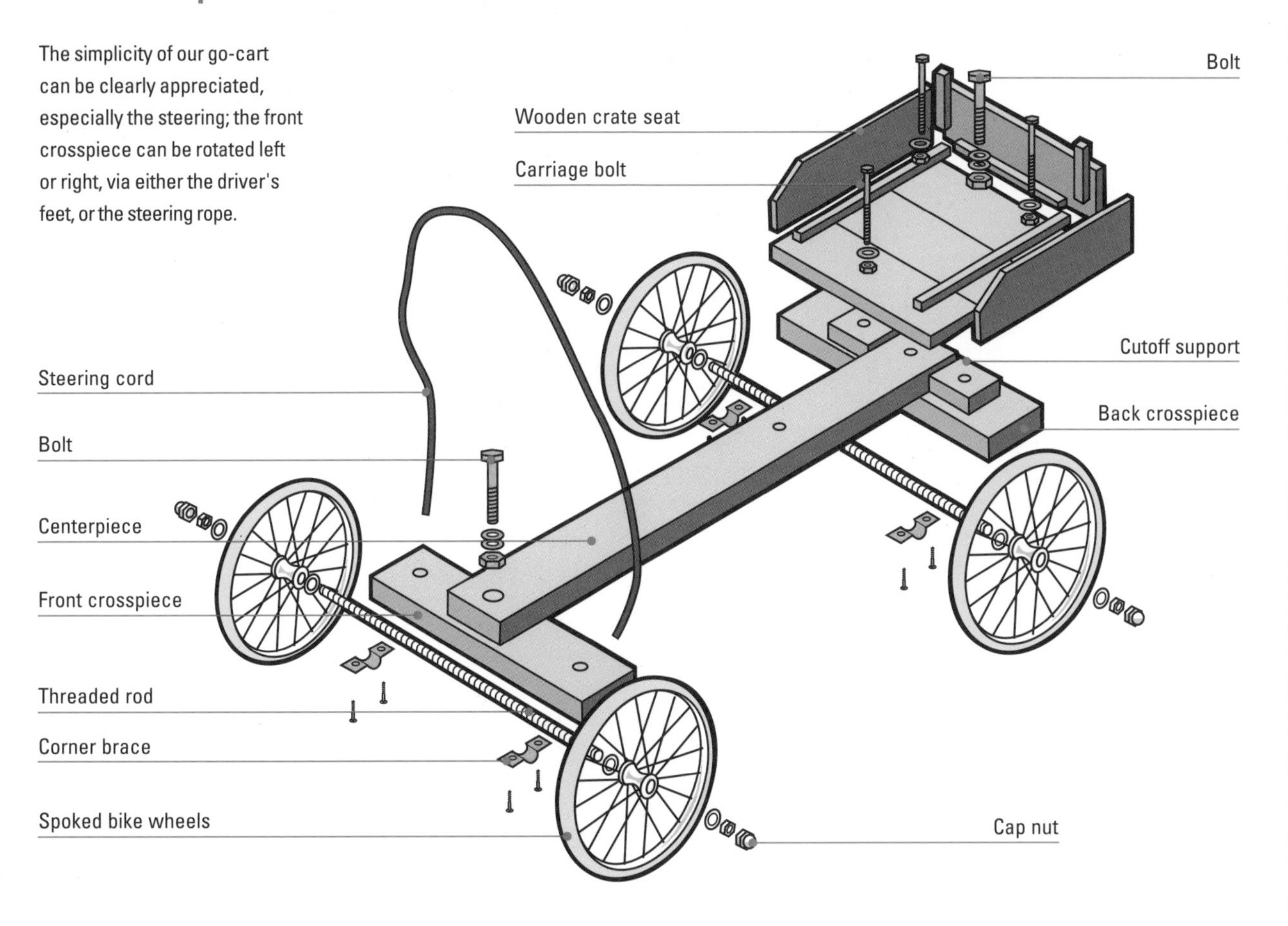

How to Use It

You're now ready to get outdoors and have hours of fun and thrills—all without pedal or gas power! Without brakes, this go-cart can be dangerous, so find a quiet grass park, ideally where there are no people or animals to run into, and use it at slow speeds only.

Riding and Steering

Our wooden go-cart is not self-propelled, which means that someone needs to push it and the driver, or simply find a gentle hill to ride down. The steering mechanism is very simple: the front crosspiece can be rotated left or right, and can be controlled by the driver's feet, or by using the steering cord. Your feet supply the brakes, unless of course, you want to go one better and add a braking system (using a lever to apply friction to the back wheels).

Controlling the go-cart

NERO
CZARNA
ČERNÁ
FEKETE
℮ 50ml

Boiler Piston Car

If you remember learning about the basic physics of expansion and contraction in school, the boiler piston car brings those rules to life. Stephenson's Rocket is an early inspiration as one of the first locomotives to be invented using the combination of the power of heat and water to propel it along its tracks.

While this little machine isn't the fastest mover, the heating and cooling of its integral boiler gives it a putt-putt motion as it runs that is very delightful, as is the idea that something as simple as a tea light provides its sole power. Although the chassis is remarkably light, using broad wheels will keep the running as steady as possible, while the simple can-and-syringe arrangement recharges itself.

YOU WILL NEED

- Metal cookie container lid (for cutting up for pieces and strips)
- Two 10ml plastic syringes (new ones used for filling printer cartridges)
- Shoe polish can, approximately 2¾ inches (70mm) in diameter
- Three 1-inch (25-mm) copper rivets, ¼ inch (6mm) in diameter
- Three copper roofing nails
- 39½ inches (1,000mm) of 1/16-inch (2-mm) diameter copper electrical wire
- Tea light
- Two brass tubes, ¼ inch (6mm) bore, one at 4¾ inches (120mm) long, the other at 6⅝ inches (170mm)
- 8-inch (200-mm) length of ¼-inch-(6-mm) diameter silicon tubing
- 8⅝-inch- (220-mm) length of ⅛-inch- (4-mm) diameter fiberglass rod
- 6⅝-inch (170-mm) length of ⅛-inch- (4-mm) diameter fiberglass rod
- Rivet plug
- Four plastic wheels from a toy, approximately 2¾ inches (70mm) in diameter
- Two plastic gears from a toy, about ⅝ inch (15mm) in diameter
- ⅝-inch (15-mm) fencing staple
- Box of matches

TOOLS

- Cordless drill/driver with a good selection of drill bits to fit)
- Flat file
- Hacksaw (to cut brass tubes)
- Marker pen
- Nail
- Pair of square and long-nose pliers
- Sandpaper
- Soldering iron (with cored solder and flux)
- Tin snips (to cut metal)
- Two tube pack of epoxy; one tube is the resin, the the other is the hardener
- Utility knife (to cut silicon tubes)

How to Make the Boiler Piston Car

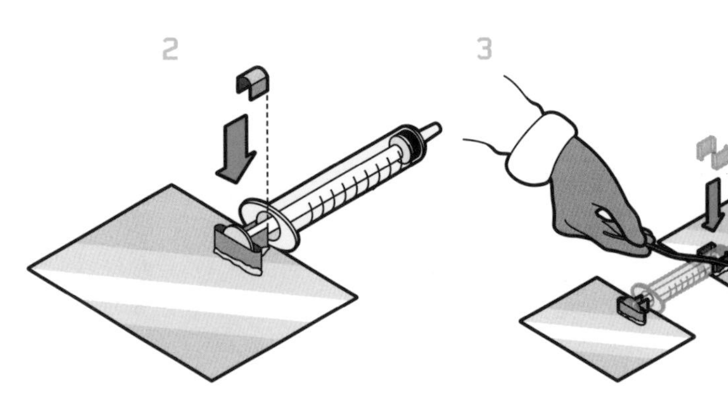

1 To make the front end, mark out and cut a piece of tin $3\frac{1}{2} \times 2\frac{3}{4}$ inches (90×70mm) using the tin snips and file the edges. Cut another strip $\frac{3}{8}$ inch (10mm) wide and fold this into a capital T-shape so that it will grip the plunger of the syringe. Solder it in the center of the long side of the plate.

2 Cut another $\frac{3}{8}$-inch- (10-mm) wide strip of tin, bend it into a bridge shape using the long-nose pliers, and solder it onto the plate so that the syringe is captive.

3 To make the back end of the car, mark up and cut a plate approximately $3\frac{1}{2} \times 2\frac{3}{4}$ inches (90×70mm) and a long $\frac{5}{8}$-inch- (15-mm) wide strip. Bend the strip using square pliers to form two small channel sections and solder these in the center of the long edge of the plate so that they grip the barrel of the syringe.

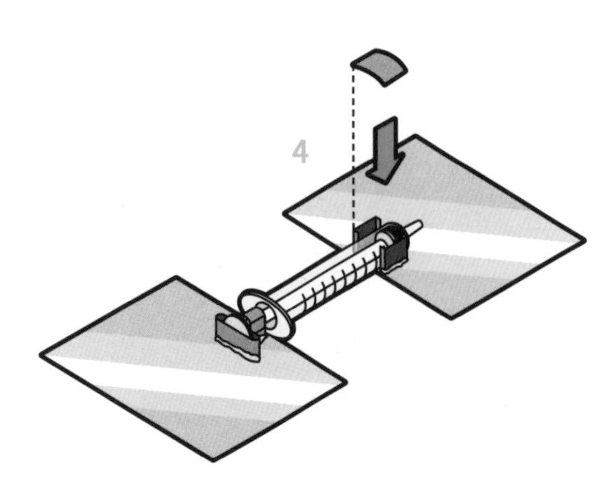

4 Using a piece of the $\frac{5}{8}$-inch- (15-mm) wide strip, solder a strap across the two channel sections to grip the syringe barrel firmly.

5 To make the boiler, use a nail to pierce a small hole in the top of the polish can. Sandpaper the rim of both halves of the pan generously around the nail hole to let the solder bond with the surface effectively.

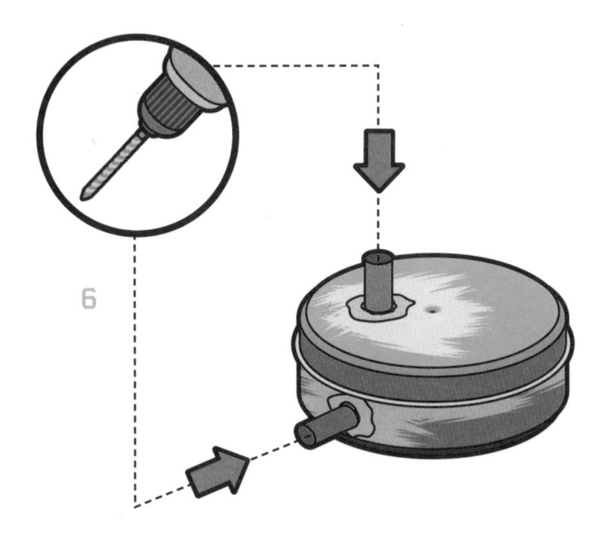

6 Drill a small 1/16-inch (2-mm) hole through the center of the rivets to make a flanged tube and solder these onto the can, as shown. Then drill through the rivet into and through the can.

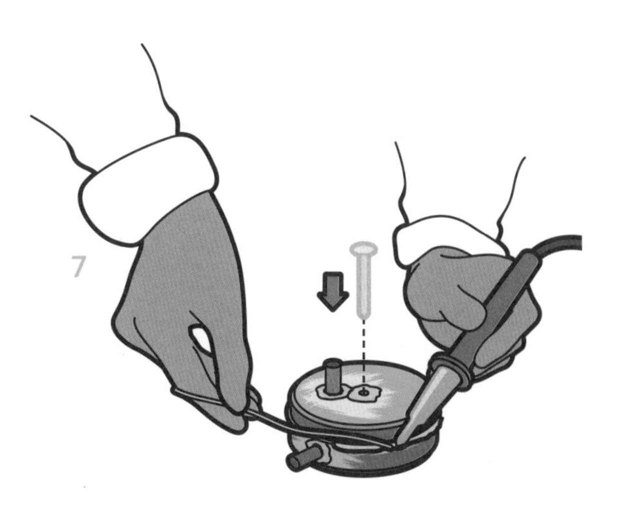

7 Using a copper roofing nail, punch a hole right through the center of the can, top and bottom. Pass the nail through the can and solder it top and bottom (this stops the can from bending in the middle). Solder around the edge of the can.

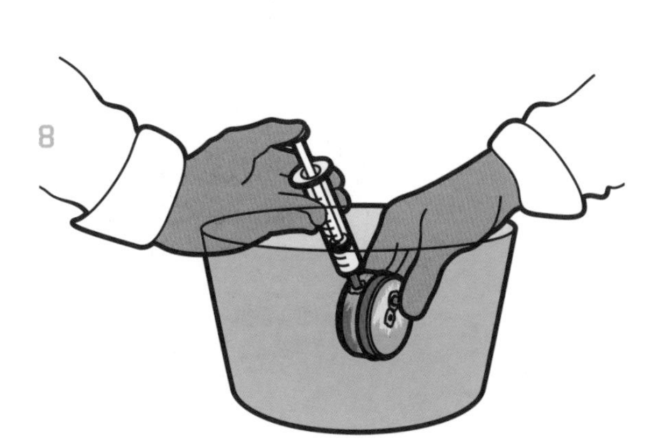

8 Immerse the can in a pan of water and use the syringe to blow air into one of the rivets, covering the other one up with your thumb to find out if there are any air leaks around the can. Fix any air leaks with solder.

9 Mount the boiler approximately 1 3/8 inches (30mm) above the base plate (to which the piston and later the wheels are mounted), soldering on the square copper electrical wire for legs.

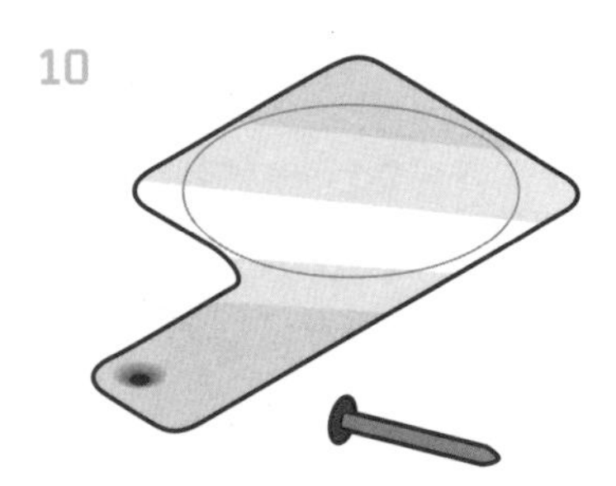

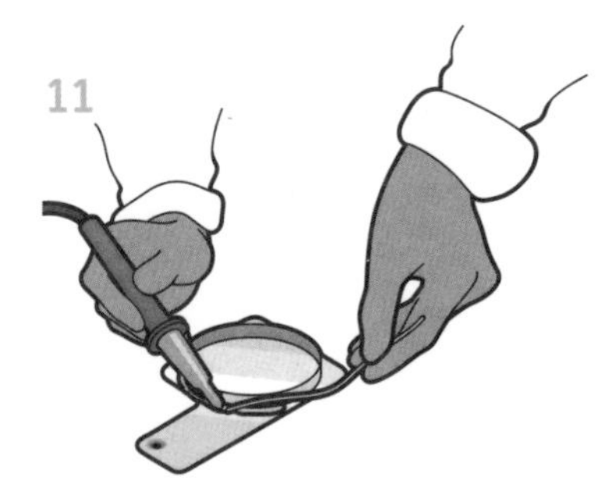

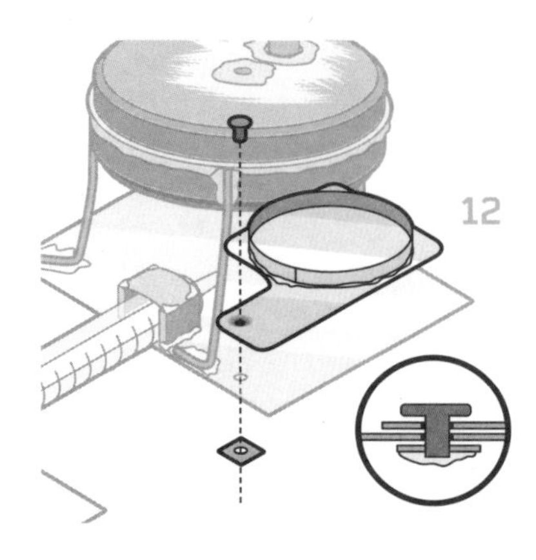

10 Mark out and cut out a piece of brass (or tin), approximately $2^1/_2 \times 1^1/_2$ inches (65 × 40mm), and trim to the shape shown in the template on page 156 making sure that a tea light will fit on the square end. In the $^5/_8$-inch- (15-mm) wide end, punch a hole using the nail.

11 Cut a $^1/_4$-inch- (5-mm) wide long strip of tin and solder it to the plate, making a rim to hold the tea light.

12 Cut short a roofing nail with a pair of pliers, drill a hole in the back plate for the nail to go through, as shown, and fasten using a $^1/_4$-inch (6-mm) square tin washer (the washer is soldered to the nail but not to the chassis to make a pivot).

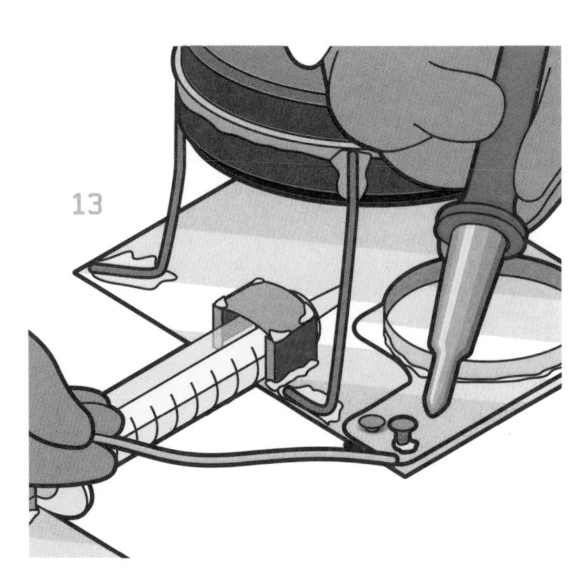

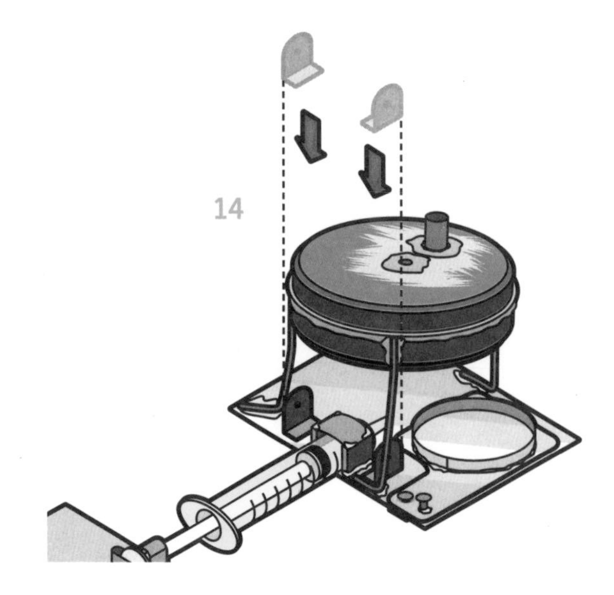

13 Cut another roofing nail with a pair of pliers so that it is only $^1/_4$ inch (6mm) long and solder it to the tea-light tray (this nail acts like a fixed post on the tea-light tray).

14 To make the brackets, cut a long strip of tin approximately $^5/_8$ inch (15mm) wide. Cut two pieces off, each 1 inch (25mm) long, and bend them to a right angle as shown. Drill the $^1/_{16}$-inch (2-mm) holes (the position of these is not critical as long as the wire can pass through them) and solder the strips to the chassis, left and right of the barrel mounting.

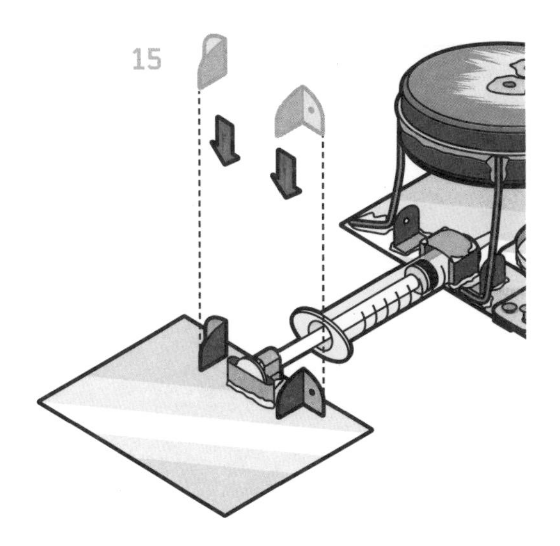

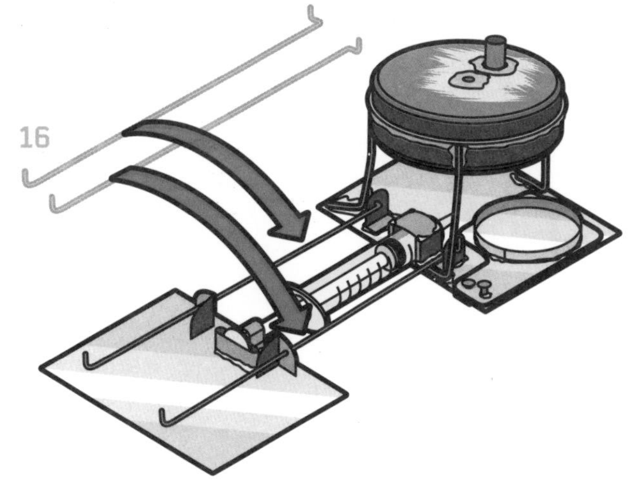

15 To make the end stops, cut two strips of tin approximately ⅝ inch (15mm) wide and 1 inch (25mm) long, and bend then to a right angle as shown (these, and the brackets in the previous step, should be as similar as possible, but do not need to be identical). Drill the ¹⁄₁₆-inch (2-mm) holes and solder the stops to the chassis.

16 Make two 5¾-inch (145-mm) stopping rods from the copper wire, bent to the pattern with long pliers. Attach the copper stop rods, as shown, to secure the front and back half of the chassis (these stop the syringe from pushing out of the barrel).

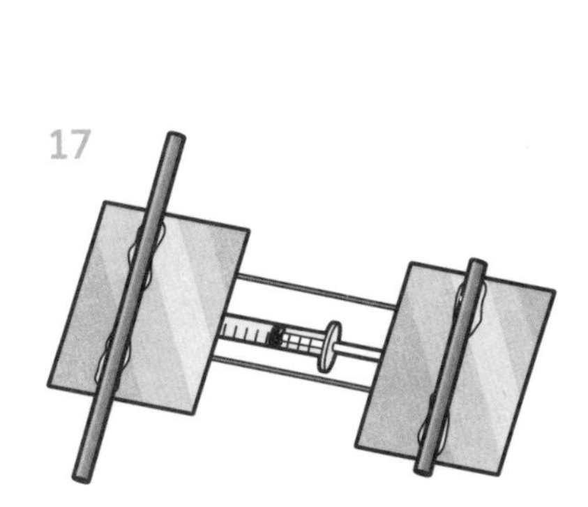

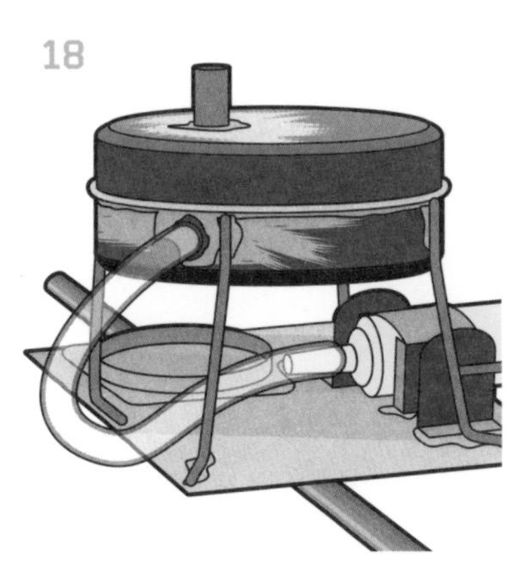

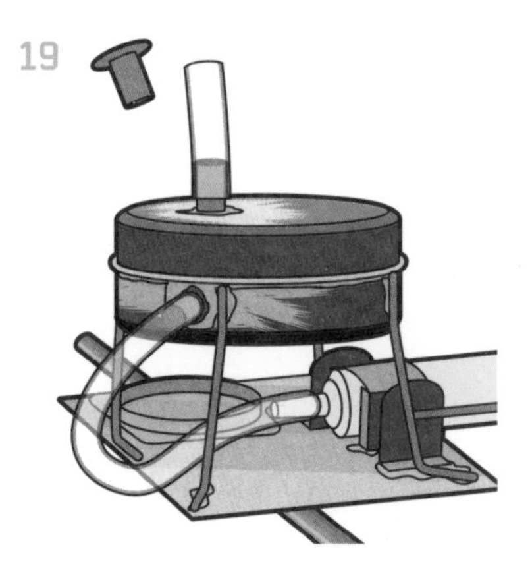

17 Solder the brass tubes to the underside of the chassis, as shown, with the shorter axis at the front of the vehicle and the longer axis at the back.

18 Secure the silicon tube from the boiler to the syringe. The silicon tube sits tightly to form a good seal.

19 Attach the silicon tube from the boiler filler to the rivet plug. The rivet plug is placed inside the silicon tube to close it off when required.

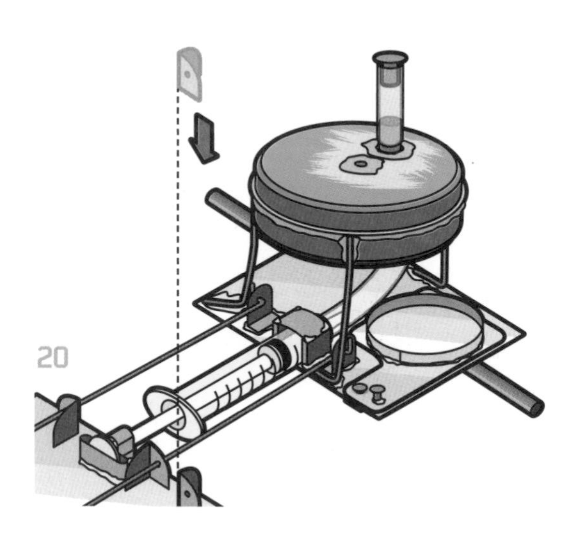

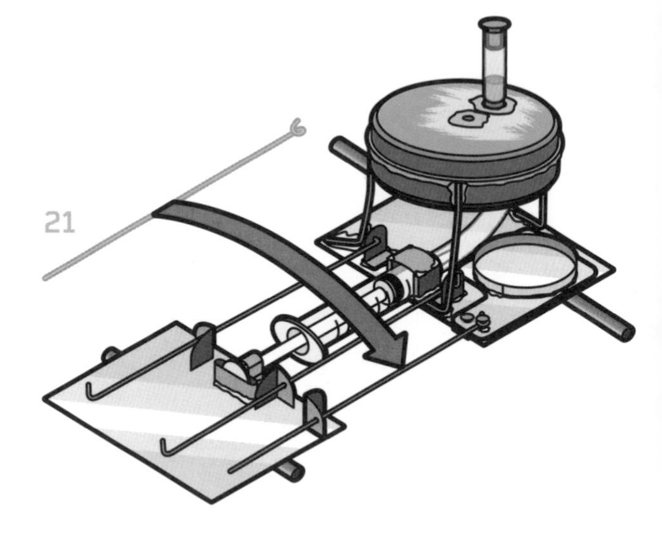

20 Use another right-angled 1-inch (25-mm) piece of tin to make a bracket on the front of the chassis (this is used to pull the tea light out from under the boiler as the car extends).

21 Use a piece of copper wire to make a link from the tea-light tray to the front bracket.

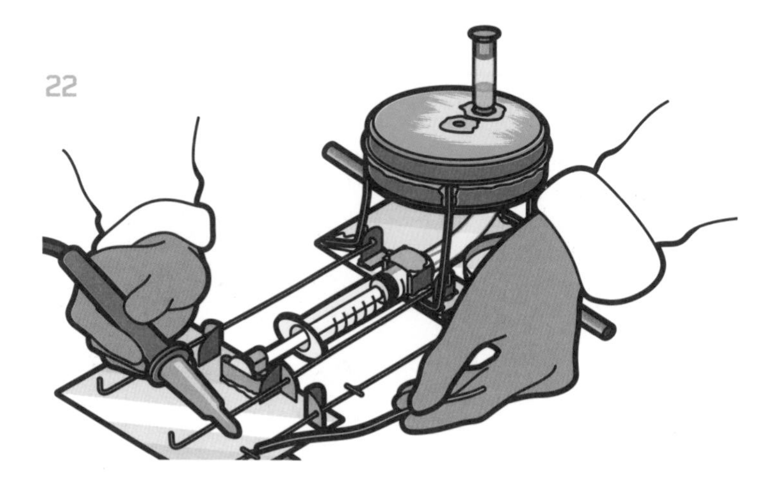

22 Solder two ¼-inch- (6-mm) long scraps of copper wire across this rod, 3 inches (75mm) and 5 inches (125mm) from the tea-light end. This may well need to be adjusted during testing.

23 Attach the fiberglass rods to one each of the front and back wheels, using epoxy resin. Using resin, attach the two gears on the axle rods ⅜ inch (10mm) inboard of the wheels.

24

25

26

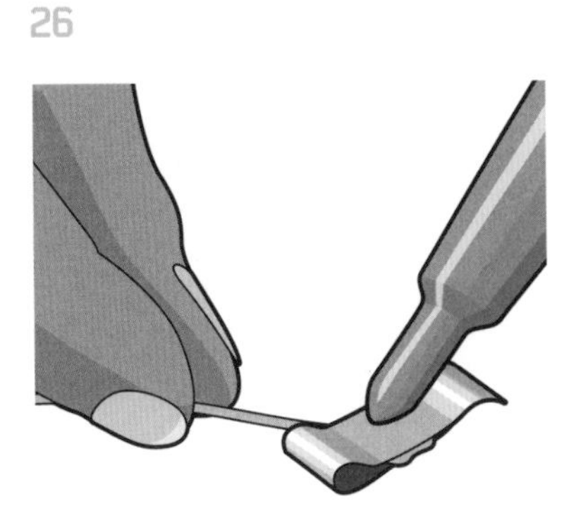

24 Pass the axles through the brass tubes on the chassis and apply resin to the wheel on the other side.

25 Bend a ⅝-inch (15-mm) fencing staple to a Z shape, as shown, and solder it to the front of the vehicle over the gear wheel.

26 Make a ratchet using a 1 inch × ¼ inch (25 × 7mm) piece of tin folded in half and soldered together.

27

28

27 Locate the pawl (ratchet) on the staple, and solder a washer of scrap tin to the end of the staple to hold the ratchet on.

28 Repeat steps 25–27 for the back end, this time attaching the ratchet to the axle tube (ratchets ensure that both sets of wheels can travel in only a forward direction).

Boiler Piston Car

Our car moves forward demonstrating the simple physics of expansion and contraction. As air in the boiler heats, it expands, pushing the syringe out and pushing forward the front wheels. As the air cools and contracts, the air is drawn back in.

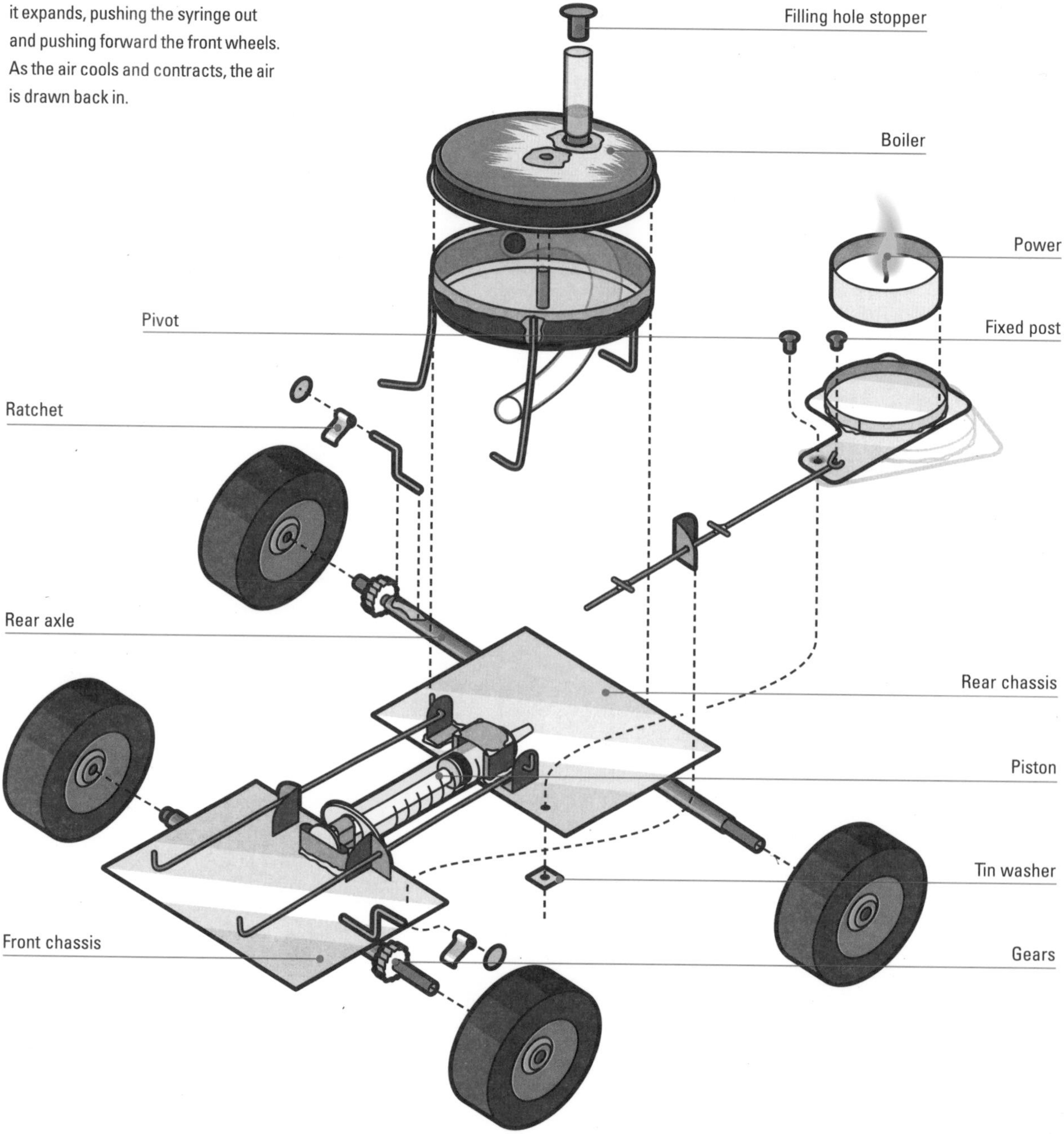

How to Use It

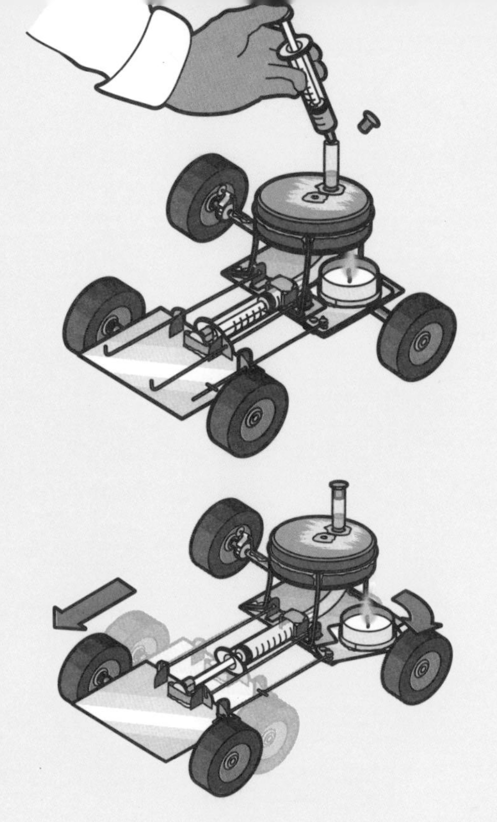

You're almost ready to set up the the car; all you need to do is to prepare the syringe with water for the boiler and light the tea light to set the car off. Be patient, because you will need to wait for the boiler to heat up first. You may also need to make some adjustments to the car on first testing.

Firing up the Boiler

Put half a teaspoon (2ml) of water into the filling tube. With the vehicule in the closed position, light the tea light.

Movement of the Car

When the top of the boiler is hot to touch, insert the plug into the filling tube. After a few minutes, the water will heat up and push the syringe out, and the tea light will swivel out from under the boiler as the syringe reaches full extension. The boiler now cools, the air contracts, and the syringe is drawn back in. The vehicule moves forward extremely slowly in a crabbing motion, front wheels first, then back wheels.

⚠ Safety

- Be particularly careful when cutting the tin. Wear gloves to cut the metal. File the edges carefully and be aware they may still be sharp when handling the car.
- The car should be used outside, where there is plenty of space to avoid bumps and crashes.
- You are working with a lit flame to light the boiler, so be careful with the matches and the tea light. Stand well clear of the car.

The tea light heats up the car's boiler.

Light-Following Robot

Our robot is trained to respond to infrared light—just as a TV responds to its remote. It has two infrared sensors on its front and turns toward the transmitter when the latter is pointed at it from a distance. We chose a utilitarian look, with a loudspeaker mounted on a flat piece of medium-density fiberboard (MDF) and a wedge-shape front, but you can customize its "character" if you want. The only requirement is that the light path to the photosensors must not be obstructed by new features. Light-following robots are popular as ready-to-assemble kits, available online. Don't be tempted by these; use our template to build your own from scratch instead and make it yours with a few antennae or rolling eyes.

YOU WILL NEED

MDF to make up:

For the transmitter:

- Two pieces, $1\frac{3}{8} \times \frac{3}{8} \times \frac{1}{4}$ inch (35 × 10 × 6mm)
- Piece, $1\frac{3}{8} \times 3\frac{1}{2} \times \frac{1}{4}$ inch (35 × 90 × 6mm)

For the robot:

- A: piece, $\frac{3}{4} \times 2\frac{3}{8} \times \frac{1}{4}$ inch (18 × 60 × 6mm)
- B: piece, $\frac{3}{4} \times 2\frac{3}{8} \times \frac{1}{4}$ inch (18 × 60 × 6mm)
- C: piece, $3\frac{3}{4} \times 4\frac{11}{16} \times 1\frac{1}{4}$ inch (94 × 117 × 6mm)
- D: piece, $3\frac{3}{4} \times 3 \times \frac{1}{4}$ inch (94 × 77 × 6mm)
- E: piece, $1\frac{1}{4} \times 3 \times \frac{1}{4}$ inch (33 × 77 × 6mm)
- F: piece, $1\frac{1}{4} \times 3 \times \frac{1}{4}$ inch (33 × 77 × 6mm)
- G: piece, $1\frac{1}{8} \times 3 \times \frac{1}{4}$ inch (30 × 77 × 6mm)
- H: piece, $1\frac{1}{8} \times 3 \times \frac{1}{4}$ inch (30 × 77 × 6mm)
- K: piece, $4\frac{1}{4} \times 1\frac{3}{4} \times \frac{1}{4}$ inch (110 × 44 × 6mm)
- I: piece, $2\frac{1}{2} \times 1\frac{3}{8} \times \frac{1}{2}$ inch (62 × 36 × 12mm)
- J: piece, $3\frac{1}{2} \times \frac{3}{4} \times \frac{1}{2}$ inch (90 × 18 × 12mm)
- L: piece, $2\frac{1}{2} \times 1\frac{3}{8} \times \frac{1}{2}$ inch (66 × 34 × 12mm)

- $\frac{1}{2}$-inch (12-mm) no. 4 wood screws
- Single pole, push to make switch (SW1)
- $\frac{1}{2}$-inch (12-mm) length of 1-inch- (25-mm) diameter dowel (for the rear wheel)
- 2-inch (50-mm) length of $\frac{1}{16}$-inch- (2.5-mm) diameter brass or steel rod
- Sheet of aluminum foil, 4 × 4 inches (100 × 100mm)
- Multipurpose adhesive (to attach aluminum foil)
- 1-inch (25-mm) no. 4 wood screws

Electronic components:

For the transmitter:

- 0.1-inch- (2.54-mm) pitch matrix board, $1\frac{3}{8} \times 3\frac{1}{8}$ inch (35mm × 80mm)
- 10K ¼ W resistor (R1)
- 22R ¼ W resistor (R2)
- 2K2 ¼ W resistor (R3)
- 1K0 ¼ W resistor (R4)
- 47Nf (C1)
- 4.7nF (C2)
- 22uf 50v, Electrolytic (C3)
- BC548, NPN transistor (Q1)
- SFH 4505, Osram IR emitter (D1)
- 4011, Quad 2 IP NAND gate
- Battery holder (for two AA batteries)

For the robot:

- 0.1-inch- (2.54-mm) pitch matrix board, $4 \times 2\frac{1}{2}$ inches (50 × 38mm)
- 470R ¼ W resistor (R1)
- 470K ¼ W resistor (R2 R6 R10 R11)
- 100R ¼ W resistor (R3)
- 470R ¼ W resistor (R4 R9)
- 2K2 ¼ W resistor (R6 R11)
- 3R3 ¾ W resistor, made from 3 × 10R in parallel (R7 R12)
- 100K ¼ W resistor (R4 R9)
- 10K ¼ W resistor (R15 R16)
- 10R ¼ W resistor (R17)
- 470pF capacitor, ceramic (C1 C8)
- 100nF capacitor, ceramic (C3 C5 C9 C11)
- 22 uF 16v capacitor, electrolytic (C4 C10)
- 470 uF 16v capacitor, electrolytic (C6 C12)
- PD410PI, Sharp IR photodiode (D1 D2)
- 1N4001 Diode (D3 D4)
- Dual op. amp. (LM358)
- 8 ohm 50mm speaker (LS)
- Battery holder (4 × AA)
- ON/OFF switch, S.P.S.T
- Pololu BCM 120:1, motor gearbox (M1 M3)
- Two Pololu BCM wheels, 4611-015

TOOLS

- Cordless drill/driver with a good selection of drill bits to fit
- Hacksaw
- Hole cutter for diameter of 1½-inch (40-mm) hole, (optional, to drill the loudspeaker aperture)
- Sandpaper and sanding block
- Soldering iron (with cored solder and flux)
- Two tube pack of epoxy adhesive; one tube is the resin, the other is the hardener
- White glue

How to Make the Robot

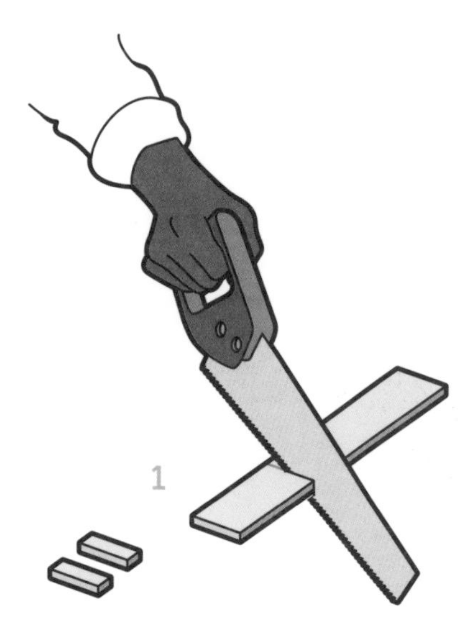

1 To make the transmitter, cut two 1⅜ × ⅜-inch (35 × 10-mm) pieces of ¼-inch (6-mm) MDF and one 1⅜ × 3½-inch (35 × 90-mm) piece to serve as the base plate. Glue the two 1⅜ × ⅜-inch (35 × 10-mm) pieces to the base plate.

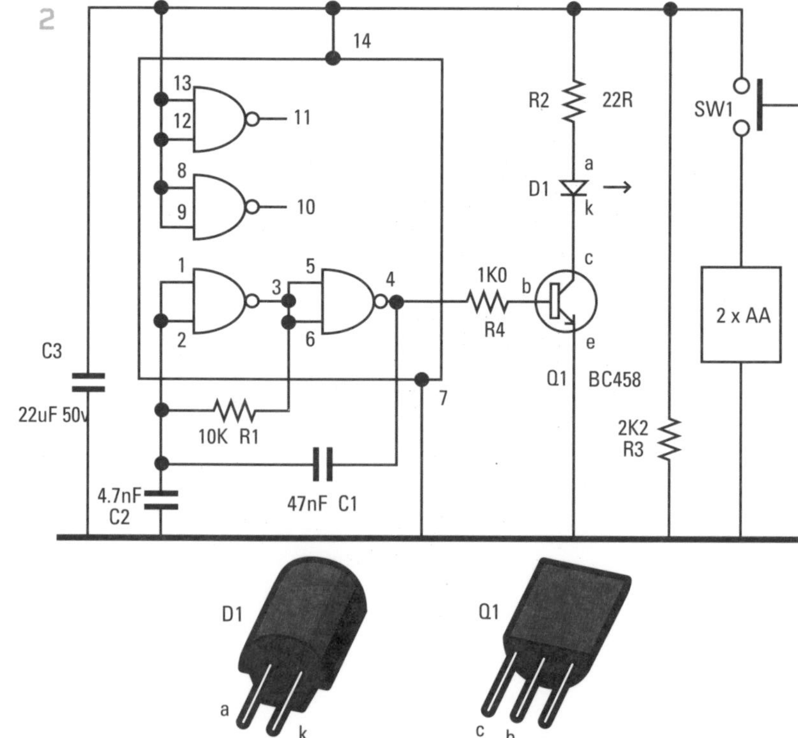

2 Arrange and wire the following parts on the matrix board, (layout is not critical), as in the circuit diagram above. Note the pinouts—the wires, or pins—for semiconductors D1 and Q1 and their position in the matrix board.

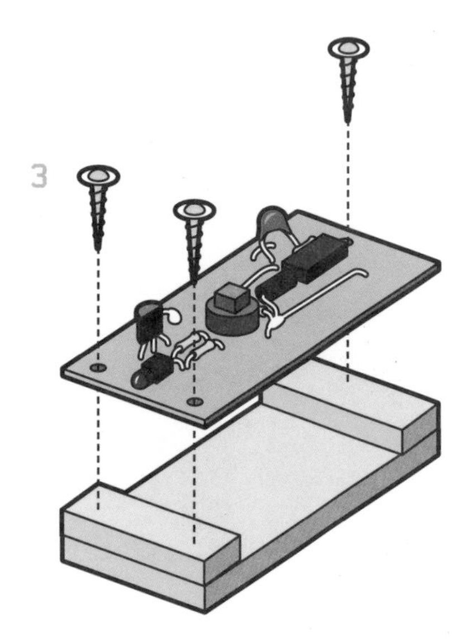

3 Fasten the finished circuit board to the base plate using the ½-inch (12-mm) no. 4 wood screws, as shown.

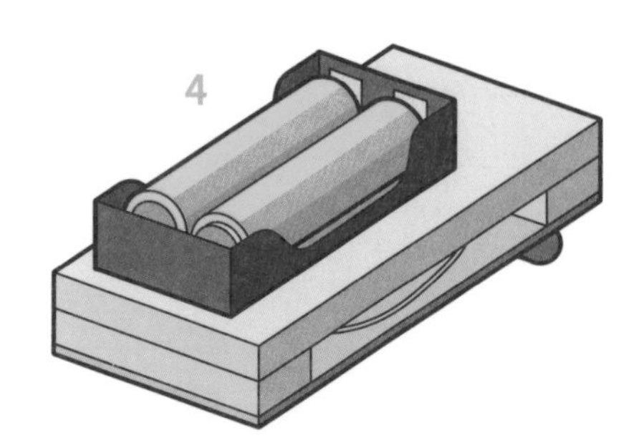

4 Turn over and attach the battery holder to the base plate with a single ½-inch (12-mm) no. 4 wood screw and wire the terminals directly to the circuit board, passing the wires through the ¼-inch (6-mm) hole at the end of the battery holder.

5 To make the robot, assemble the circuit board as shown in the circuit diagram below, with the components as indicated on the right. The location of wiring is not critical, because no high-frequency elements exist in the circuit. Some components around the IR sensor photodiodes are mounted separately on a small board located at the front of the machine (see step 14).

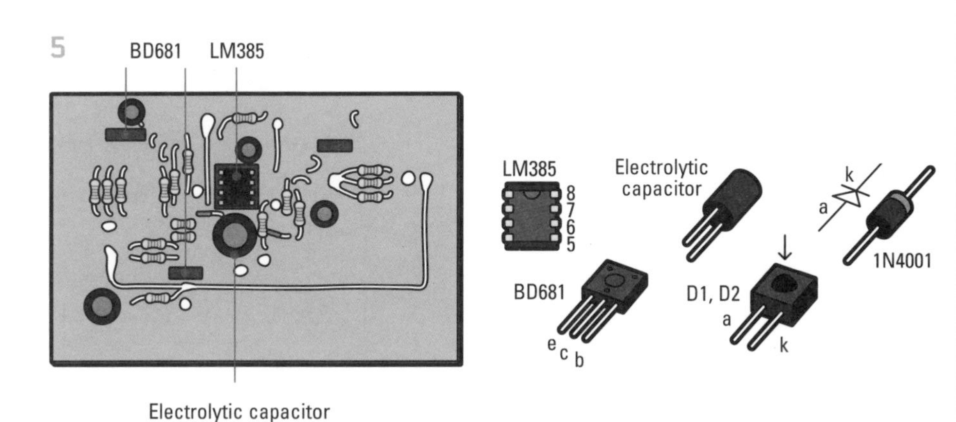

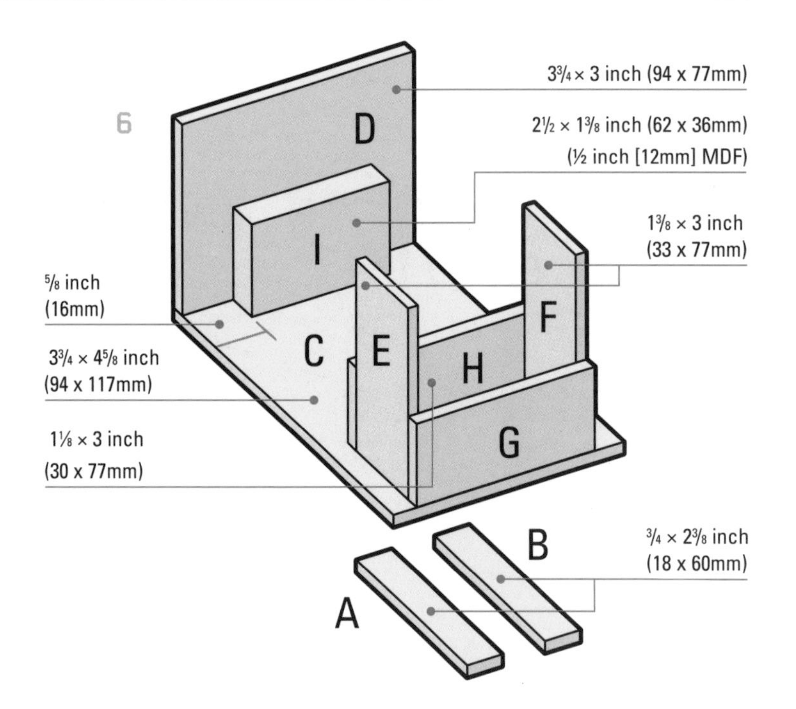

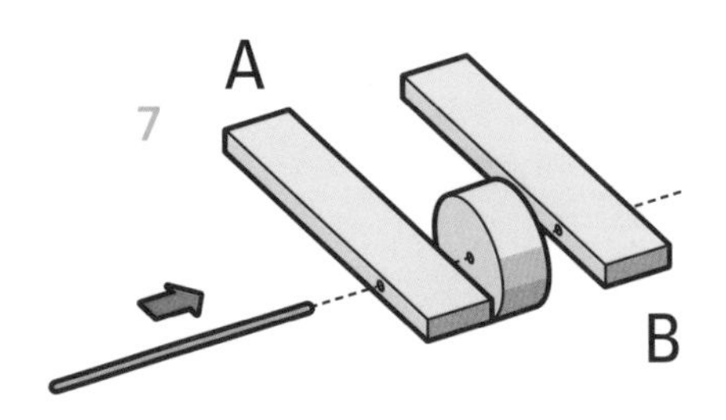

6 Using the handsaw, cut the following pieces of ¼-inch (6-mm) MDF: pieces A and B, ¾ × 2⅜ inch (18 × 60mm); piece C, 3¾ × 4⅝ inch (94 × 117mm); piece D, 3¾ × 3 inch (94 × 77mm); pieces E and F, 1⅜ × 3 inch (33 × 77mm); pieces G and H, 1⅛ × 3 inch (30 × 77mm). Again using the handsaw, cut a block of ½-inch (12-mm) MDF, into a 2½ × 1⅜ inch (62 × 36mm) piece for piece I.

7 Saw off a ½-inch (12-mm) piece of the dowel and drill a ⅛-inch (3-mm) hole through the center. Drill an ⅛-inch (3-mm) hole through the ends of pieces A and B, as shown, ½ inch (12mm) from one end. To complete the rear wheel assembly, insert the brass rod (which serves as an axle) through the holes in pieces A, B, and the wooden dowel and use epoxy to glue the rod into place.

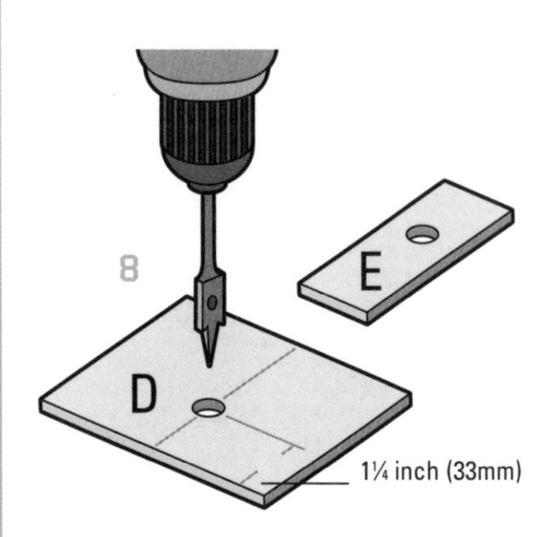

8 Drill a ⅜-inch (10-mm) hole in piece D, 1¼ inch (33mm) below the center of one of the 3-inch (77-mm) sides. Drill a similar hole in piece E (size and location to fit the ON/OFF switch).

9 Using white glue, attach piece D to one end of piece C, as shown, so that the two fit flush together. Glue block I centered against block D. Glue piece G centered to the other end of piece C. Glue pieces E and F end up against piece G, ensuring the hole for the ON/OFF switch is correctly positioned. Glue pieces E and F end up against piece G, ensuring the hole for the ON/OFF switch is the correct position. Glue piece H centered on pieces E and F.

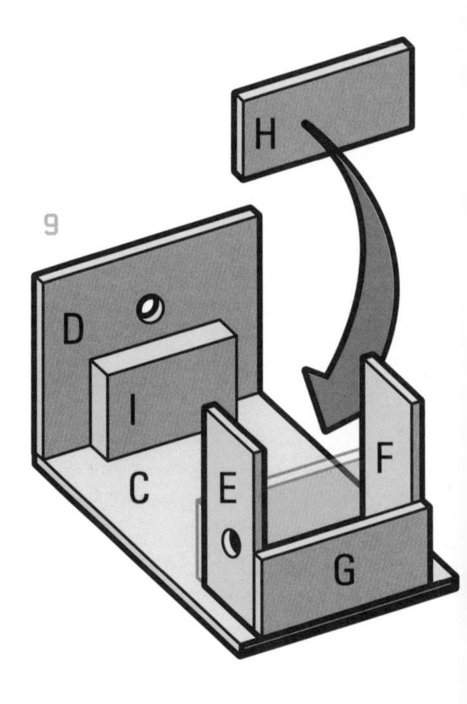

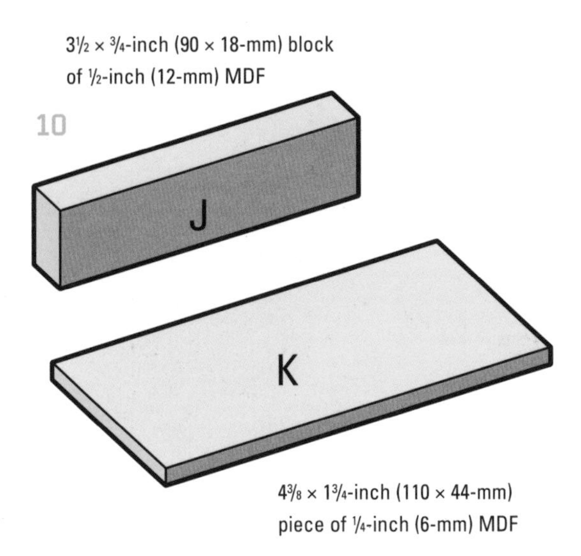

10 To make the mounts for the reflector and sensors (D1, D2), cut a 3½ × ¾-inch (90 × 18-mm) block of ½-inch (12-mm) MDF (piece J) and a 4⅜ × 1¾-inch (110 × 44-mm) piece of ¼-inch (6-mm) MDF (piece K).

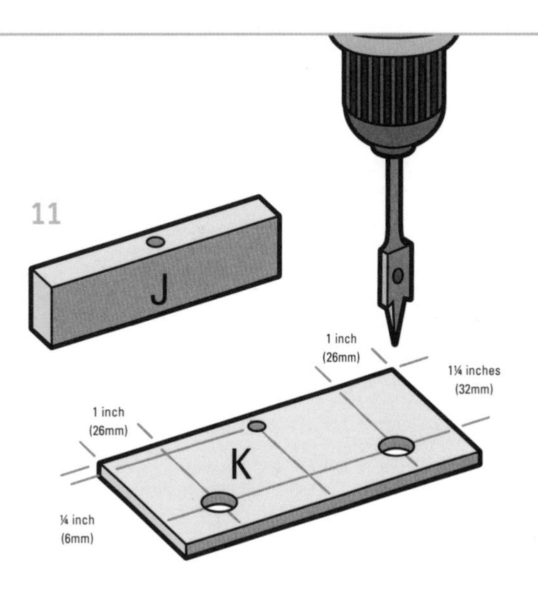

11 Drill a centered ¼-inch (6-mm) hole (for the sensor leads) through the 3½ × ½-inch (90 × 12-mm) side of piece J. Drill two ⅜-inch (10-mm) holes in piece K, 1¼ × 1 inches (32 × 26mm) in from the bottom corner, lengthwise. Drill a centered ¼-inch (6-mm) hole in piece K, which should be ¼ inch (6mm) in from the long edge of K and align vertically with the hole in piece J.

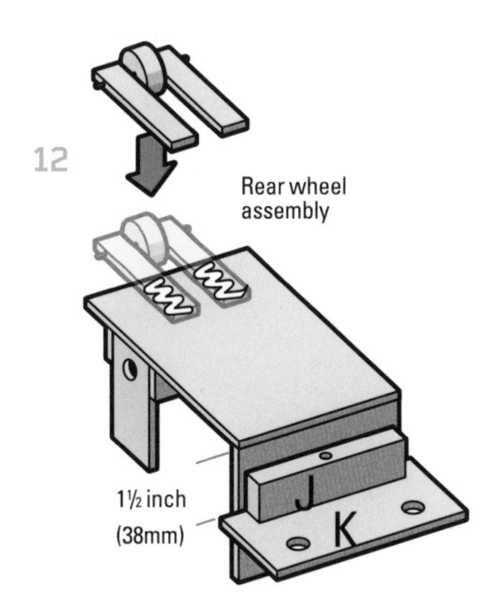

12 Glue piece K, lengthwise to piece D, 1½ inches (38mm) from the bottom, as shown. Glue block J centered beneath this. Glue the rear wheel assembly centered to the bottom of piece C at the other end of the unit, so that the wheel runs free (the front of the unit afterward should look like a wheelbarrow front).

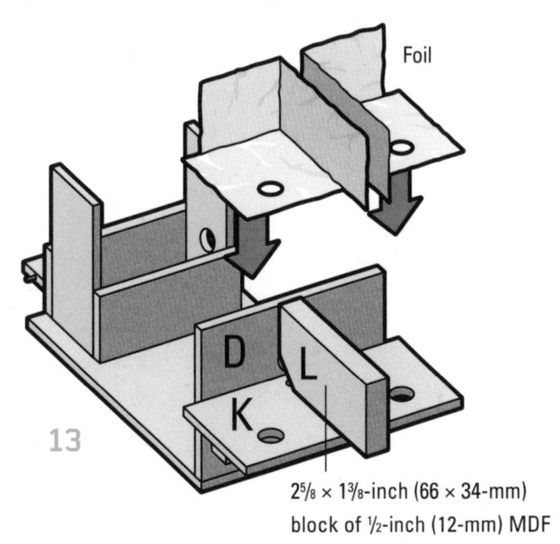

13 Cut a 2⅝ × 1⅜-inch (66 × 34-mm) block of ½-inch (12-mm) MDF (piece L) and cut a small triangle of wood from one corner. Glue L to pieces D and K, as shown. Cover the rectangular sides of piece L with aluminum foil, securing it with adhesive. Do the same with the exposed top surfaces of piece K and the exposed surfaces of piece D above this.

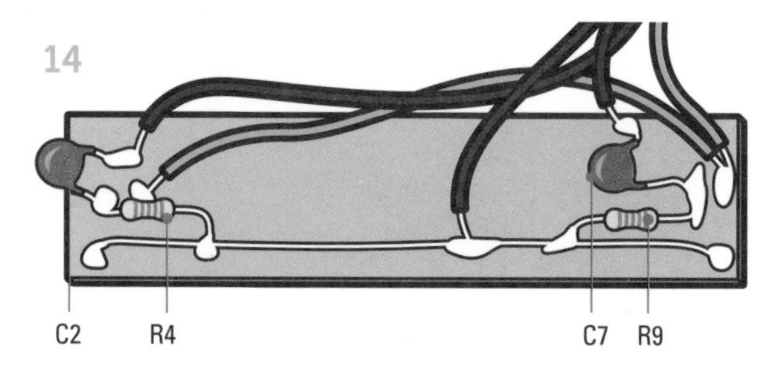

14 Solder the sensors into a piece of matrix board and, using white glue, attach the board onto the underside of piece K so that the sensors fit into the two holes, domed sides facing upward.

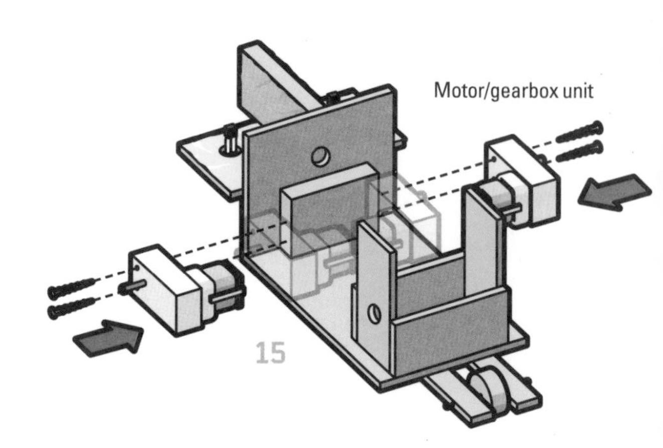

15 Mount the motor/gearbox units, securing them into position using two 1-inch (25-mm) no. 4 wood screws. It may be necessary to open up the holes on these units with a ⅛-inch (3-mm) drill bit.

16 Secure the completed circuit into position using ½-inch (12-mm) no. 4 wood screws, then connect the wires to the motor/gearbox and sensors but not yet to the loudspeaker. (The setting-up procedure may require that wires to the loudspeaker are changed until the correct configuration has been achieved, so this should be connected by a pair of long wires, which will be trimmed to the correct length when testing is complete.)

Battery unit

Circuit board

16

Photosensor

Motor/gearbox unit

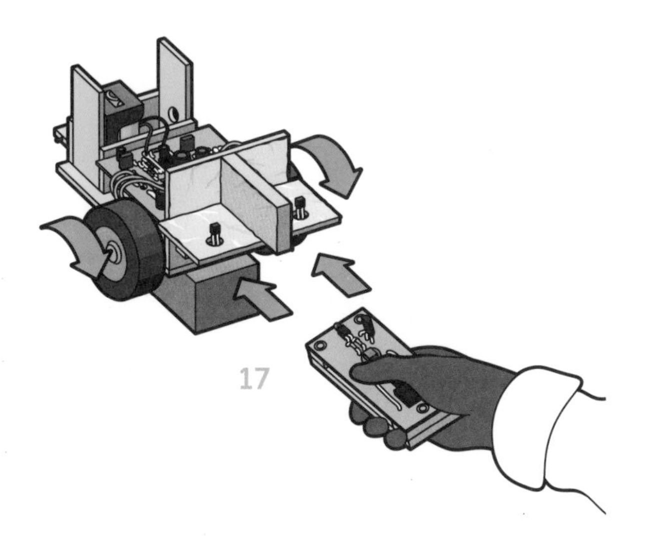

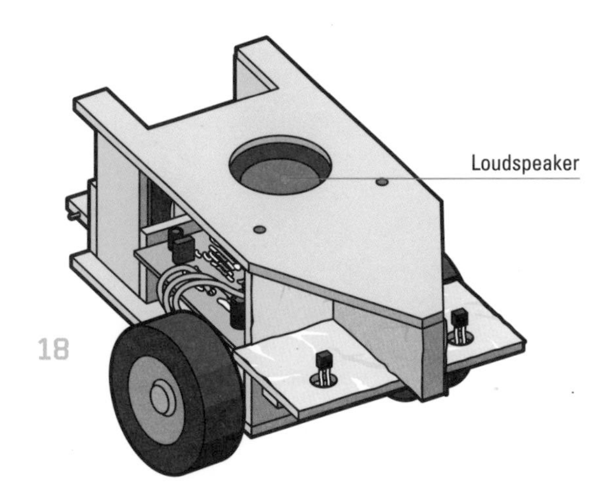

17 Slide the wheels onto the motor/gearbox output shaft, which has a flat on it so that they are firmly located. Attach using the screw provided with the wheel, then set the machine on a block of wood (to keep the wheels from touching the ground), connect the batteries, point the transmitter at the front of the machine, and check that the wheels both rotate in the direction to drive the robot forward (a 1KHz note should also be heard). If not, reverse the wires, as necessary.

18 Customize the machine to taste. Here, the loudspeaker has been mounted on a flat sheet of MDF, secured on top of the machine with two ½-inch (12-mm) no. 4 wood screws (visible in front of the loudspeaker). The cutout piece at the rear is essential so the batteries can be inserted, but the wedge-shape part at the front is just "styling." As long as the light path to the photosensors is not obstructed, the final design is up to the builder.

How to Use It

Find a clear area to use your robot. It will receive an equal strength of signal when the transmitter is placed directly in front of it. You can control the sensitivity of the steering via the strength of the infrared from the transmitter on each of its sensors.

Forward, Left, and Right

Hold the transmitter at a distance of 20 inches (500mm) in front of the robot. The sensors will receive an infrared signal, which generates an output via an amplifier to turn on the robot's motors. At this point the sound system will emit a 1KHz note and the robot will start to move. Point at each sensor in turn. The opposite wheel will rotate, thus turning the machine in the direction of the strongest beam.

Back view of the robot, showing the single wheel.

AAA
1.5V

Stomper

Meet Stomper. Actually, its movement is closer to an insect-style flail than a purposeful march. Powered by very simple electronics, it will walk around waving its legs wildly. Stomper is a B.E.A.M. (Biology, Electronics, Aesthectics, and Mechanics) robot. These robots can be made from simple electronic components and don't need any computers to control them—Stomper is turned on by means of a tiny switch. Small B.E.A.M. robots have recently become popular in kit form, but it's not really much harder to create your own from easy-to-buy bits and pieces. Although this is probably one of our more fussy projects, don't be put off by the sight of a circuit board; the only skills called for to make Stomper are simple soldering, plus the ability to follow clear instructions.

YOU WILL NEED

- Sheet of $\frac{1}{8}$-inch (3-mm) medium-density fiberboard (MDF)
- Sheet of hermoplastic (Perspex or a suitable alternative that softens with heat), $2\frac{3}{4} \times 2\frac{1}{2} \times \frac{1}{16}$ inches ($70 \times 63 \times 2$mm)
- Four $\frac{1}{2}$-inch (12-mm) no. 4 wood screws
- Two 8-inch (200-mm) lengths of $\frac{1}{16}$-inch- (2-mm) diameter brass rod
- Piece of 26-gauge (0.4-mm) brass sheet, $2\frac{3}{8}$ x $2\frac{3}{8}$ inches (60×60mm)
- Two light springs
- Double-sided adhesive tape

Electronic components

- 0.1-inch- (2.54-mm) pitch matrix circuit board, $3 \times 1\frac{1}{2}$ inches (75 x 37mm)
- 0.1-inch- (2.54-mm) pitch matrix circuit board, $\frac{5}{8} \times 2\frac{3}{8}$ inches (15×60mm)
- Two battery holders for AAA batteries (the type of holder with solder tags, not press connectors and a mating PP3 clip)
- Two miniature model aircraft servo motors (for front and rear motors)
- 100K $\frac{1}{4}$ W resistor (R1, R2)
- 1K $\frac{1}{4}$ W resistor (R3)
- 470R $\frac{1}{4}$ W resistor (R4, R5)
- Four 22R $\frac{1}{4}$ W resistors (R6)
- 470K preset (speed control) (RV1 RV2)
- 4.7uF 16 v (electrolytic capacitor) (C1)
- 0.22 uF 63v (ceramic capacitor) (C2)
- MC33201 (operational amplifier) (IC1)
- 2N7000 (N ch MOSFET) (Q1)
- IRF9530NPBF (P ch MOSFET) (Q2, Q3)
- IRF520NPBF (N ch MOSFET) (Q4, Q5)
- LED, 3mm green (D1–D2)
- SPDT slide switch PCB mount (mount on matrix board) (SW1)

TOOLS

- Cordless drill/driver with a good selection of drill bits to fit.
- Crosshead screwdrivers
- Hacksaw
- Long-nosed pliers
- Side cutters
- Small soldering iron (with cored solder and flux)
- Tin snips or kitchen scissors
- Two-tube pack of epoxy adhesive; one tube is the resin, the other is the hardener

How to Make Stomper

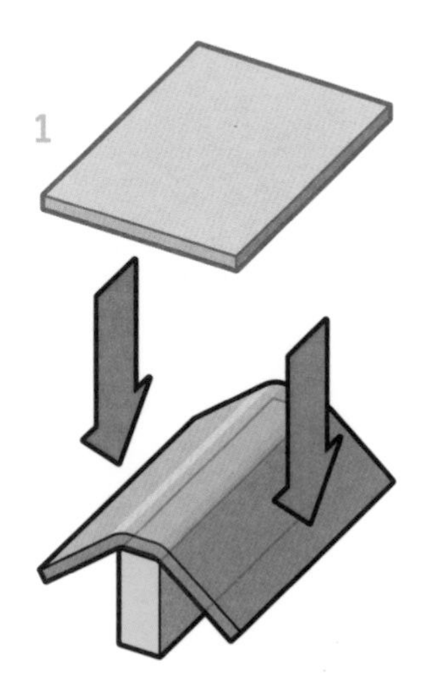

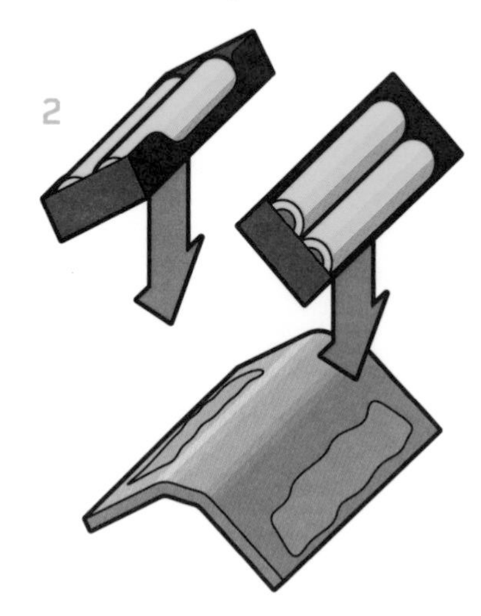

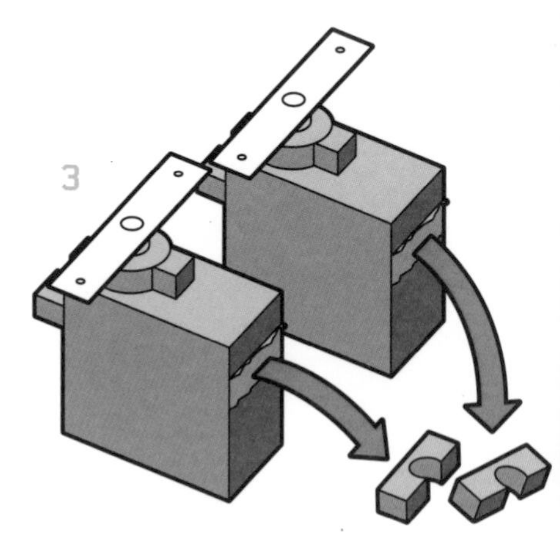

1 To make the chassis, use a piece of ½-inch (12-mm) MDF as a former, and place the plastic sheet over it in an oven at about 122°F (50°C), until it softens. When it is sufficiently flexible to be molded, fold it as shown. If it will not fold at this temperature, slowly increase the temperature in steps until the sheet becomes pliable. It may be helpful to drill two ⅛-inch (3.5-mm) holes along the top spine of the chassis material and fasten it to the MDF former with two wood screws before heating. This will make certain that it stays in the correct position while being bent. Remove the screws once the plastic sheet has been bent.

2 Use epoxy adhesive to secure the battery holders to the sloping sides of the chassis, as shown.

3 Remove the indicated lugs from the servomotors, as shown.

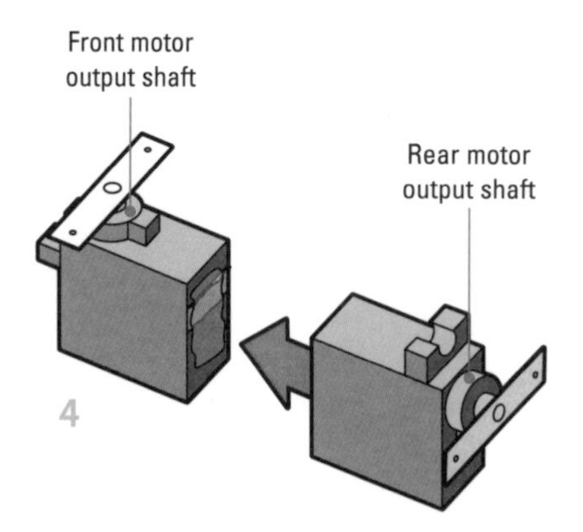

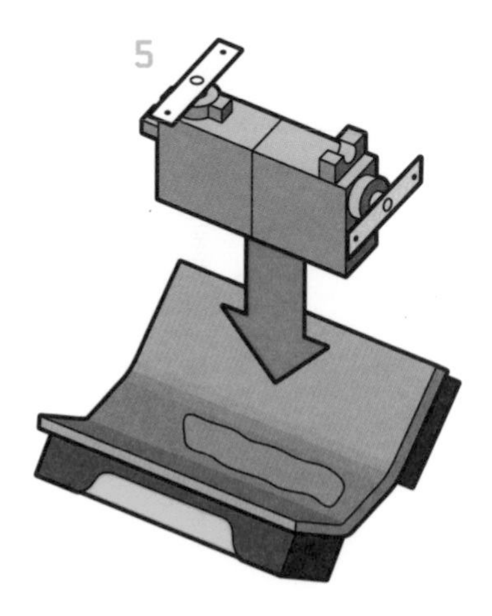

4 Glue the two motors together with the output shaft for the front motor horizontal and for the rear motor vertical. This will ensure that the front legs move up and down, while the rear legs will move side to side.

5 Attach the motors centered on the underside of the chassis with epoxy adhesive.

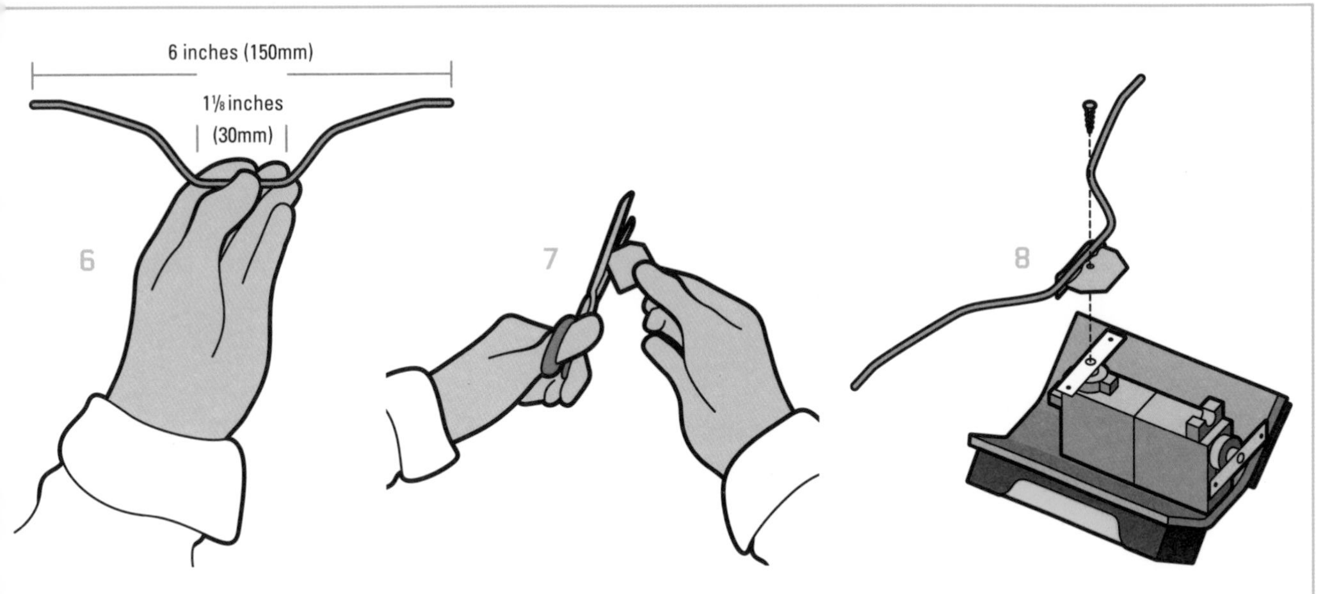

6 To make the rear legs, bend one of the lengths of brass rod into the shape shown (the exact point at which the bends are applied is not critical), so that its span, from one end to the other, is 6 inches (150mm).

7 Take the brass sheet and cut off the top and bottom edges on one long side diagonally with tin snips or kitchen scissors. Fold a ⅛-inch (3-mm) lip on the opposite side with a pair of long-nosed pliers, then drill a 1/16-inch (2-mm) hole, where shown.

8 Solder the legs to this, angling them downward. Ensure that the lip is sufficiently bent over to wrap around the white nylon actuator on the servomotor, leaving the 1/32-inch (1-mm) holes at either end uncovered. Secure the legs to the motor output shaft using the screw supplied with the motor.

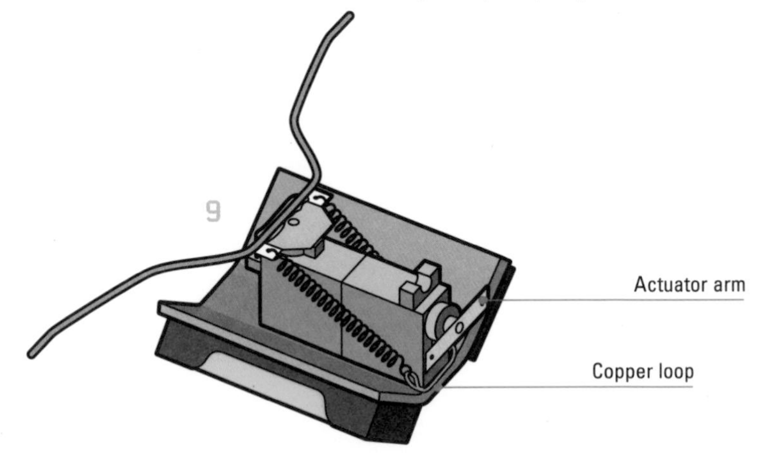

9 Hook springs in the two 1/32-inch (1-mm) holes at either end of the actuator arm. Hook the other end of each spring into a loop of copper wire, then pass this loop around the rear of the front motor. (Most hardware shops keep a box of assorted springs but these rarely have specifications attached, so the correct spring must be selected by trial and error. Those used in our machine are ¼ inch [6mm] in diameter with 20 turns.)

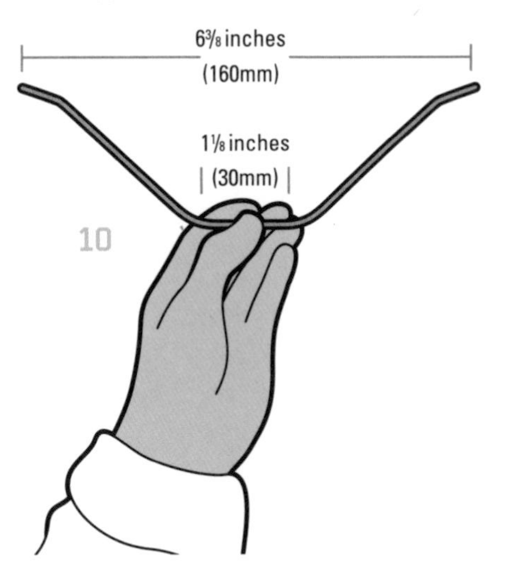

10 To make the front legs, bend the second piece of brass rod, as shown.

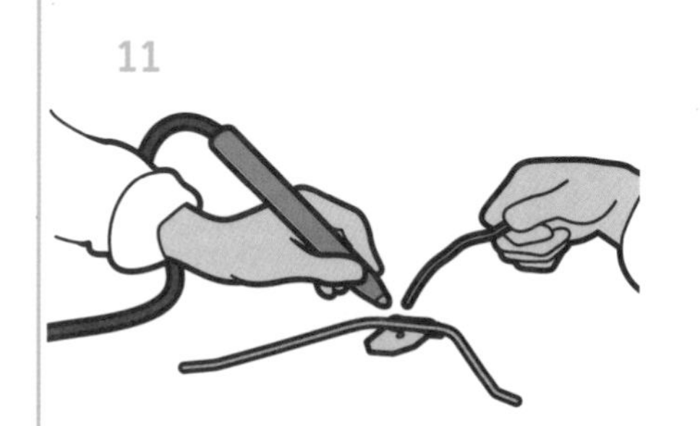

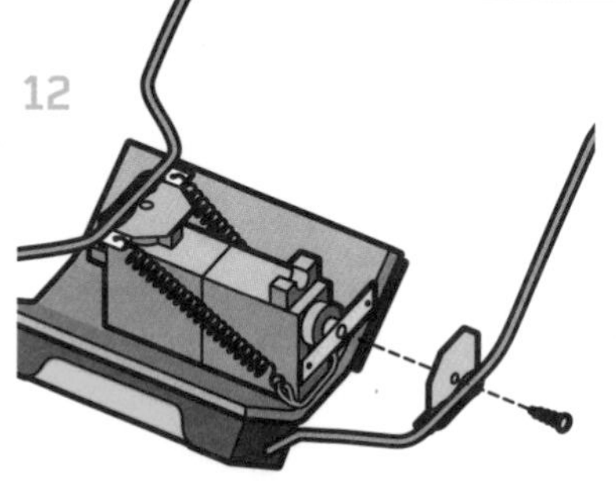

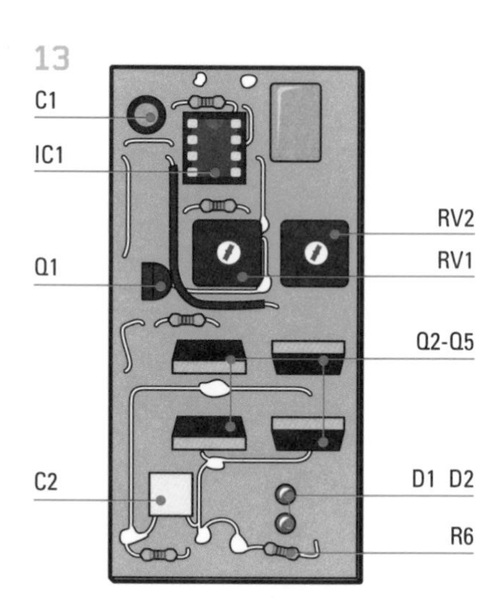

11 Use tin snips or kitchen scissors to cut a second piece of brass sheet, bend the lip with a pair of long-nosed pliers, drill a centered 1/16-inch (2-mm) hole, and then solder the front legs to this, angling them downward.

12 Screw this front leg assembly to the output shaft of the front motor, with the lip turned over to lock the brass plate to the nylon actuator. The machine should now have the general appearance of the illustration shown, minus the electronics.

13 Cut the matrix board to 3 x 1½ inches (75 x 37mm) using a hacksaw. Using side cutters to trim the components and lengths of wire, construct the electronics on the board, following the circuit diagram shown (layout is not critical).

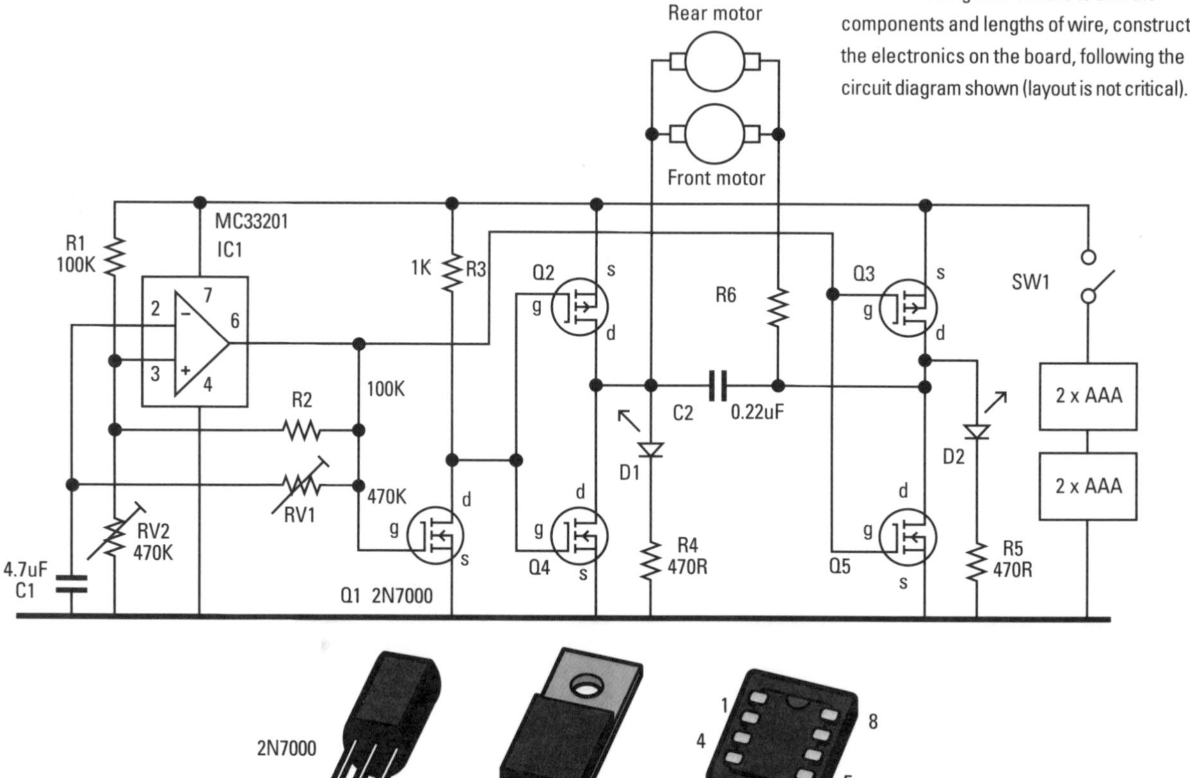

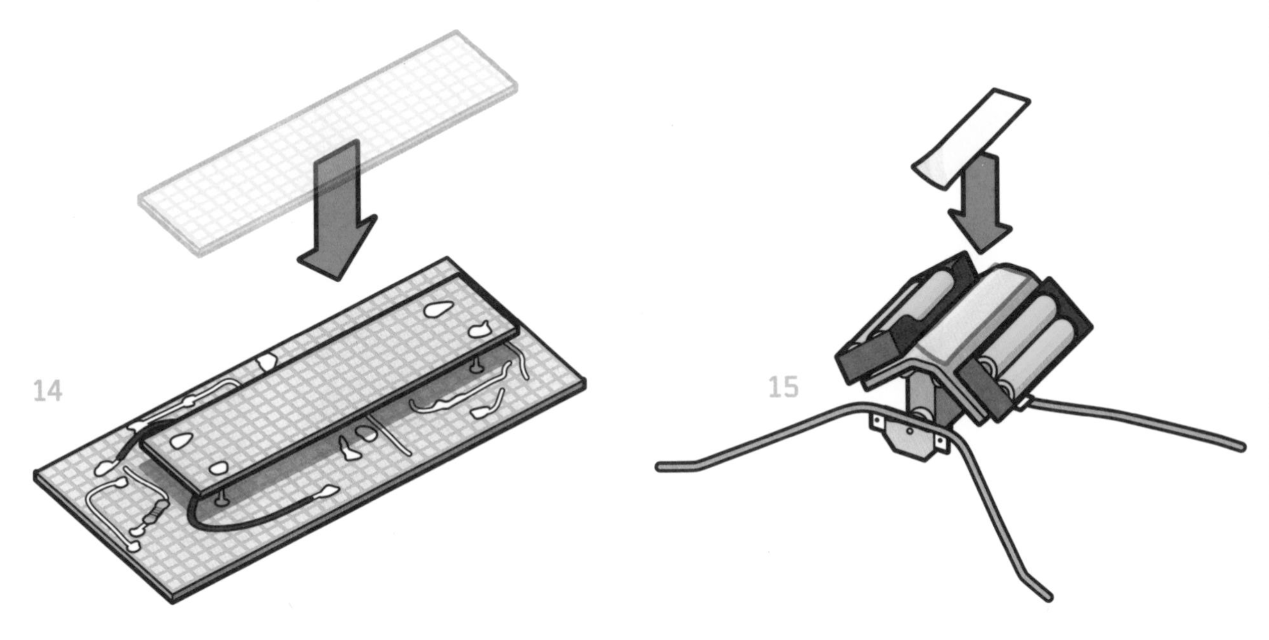

14 Solder four pieces of wire onto the bottom of the circuit board and then solder these onto the pads the second piece of matrix board (approximately ⅝ × 2⅜ inches [15 × 60mm]).

15 Place a 2⅜-inch (60-mm) length of thick double-sided adhesive tape on the flat upper surface of the chassis, as shown.

16 Push the complete circuit board assembly onto the adhesive tape, making certain that the front of the assembly is not overhanging at the front; otherwise it will be hit by the front legs when Stomper finally moves. (When complete and tested, this may be secured more firmly in place, attaching it to the sides of the battery boxes with four blobs of epoxy, one at each corner.)

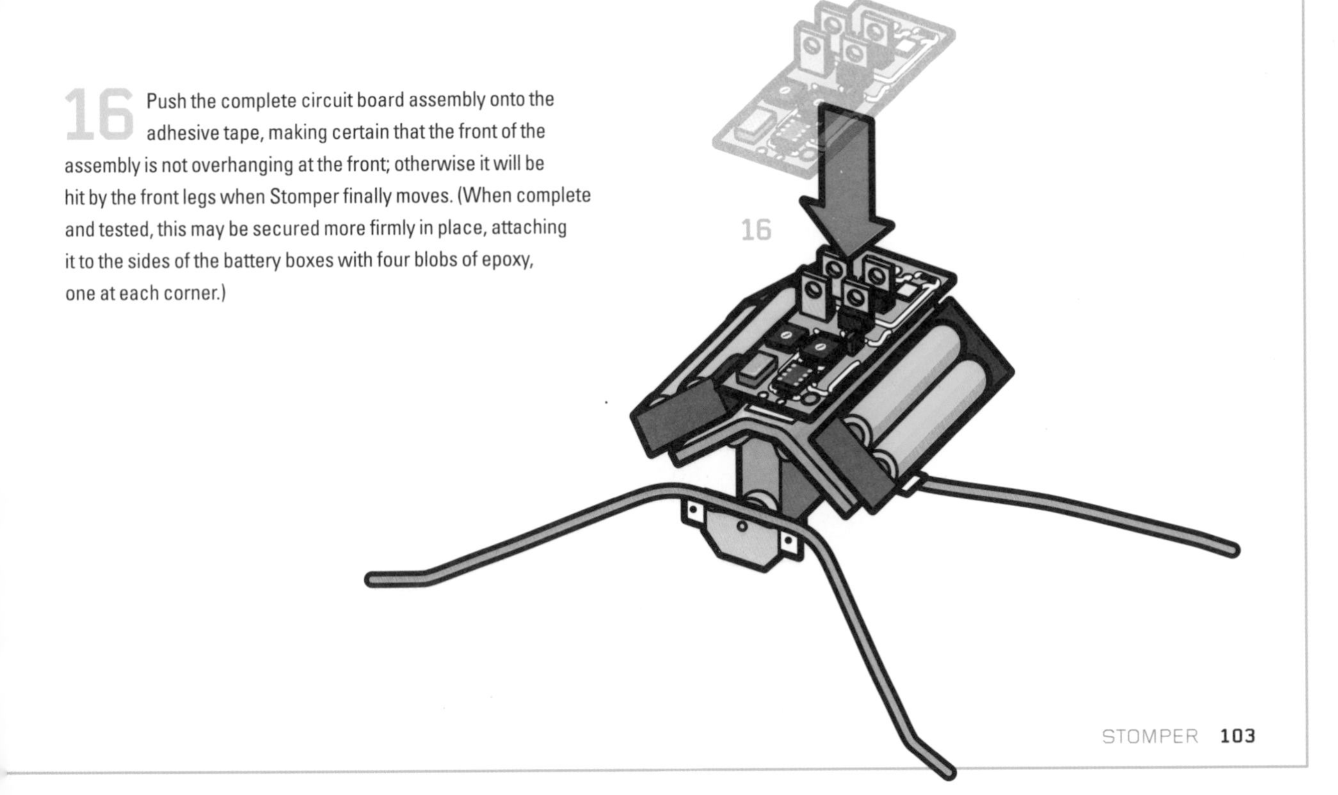

Stomper

Stomper has two independent but synchronized servomotors to drive the legs and a very simple "brain" to control its movement. The back legs provide the forward movement and the front legs raise the front of the machine on either side, thus giving the vertical lift that allows the forward "walking" to occur.

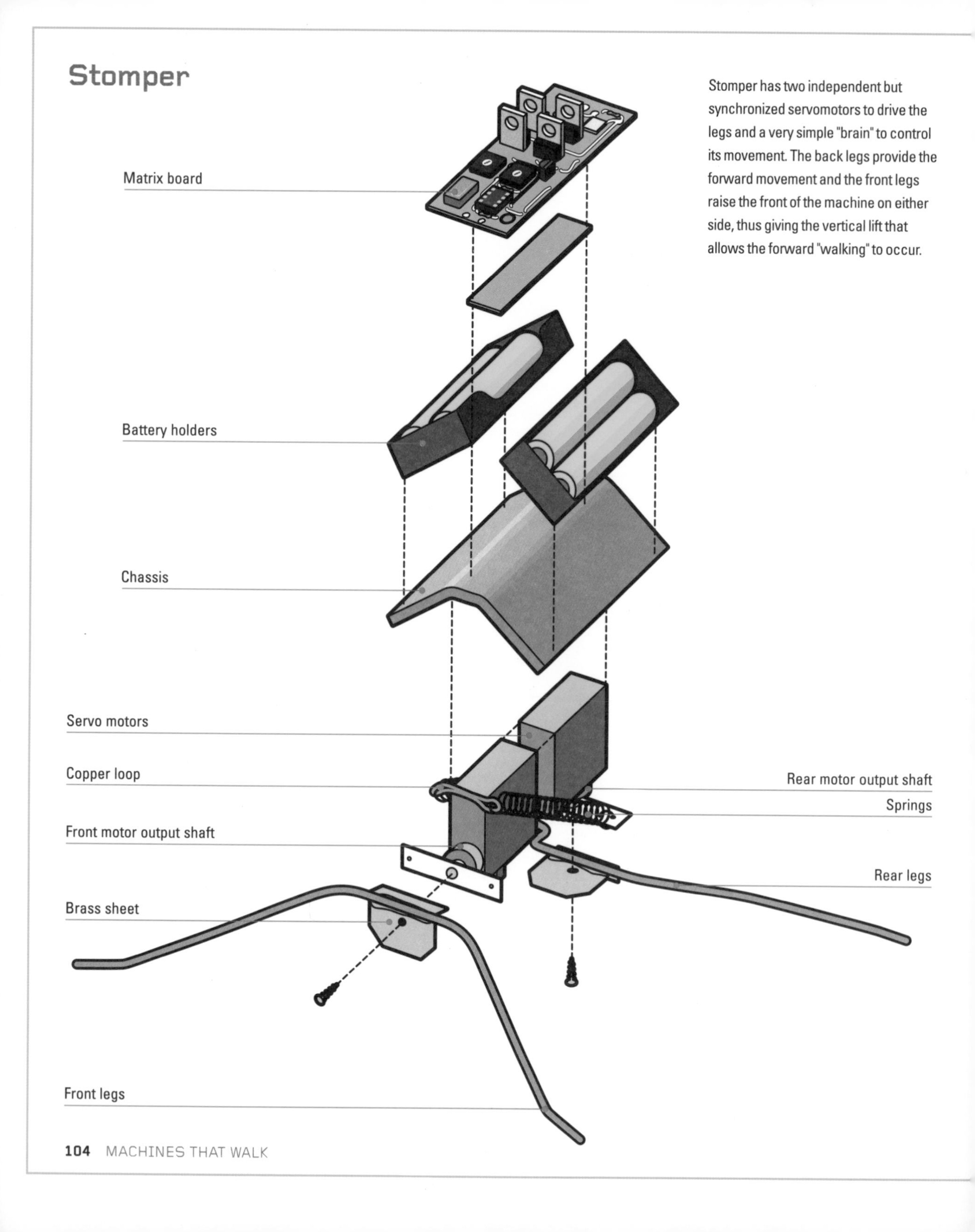

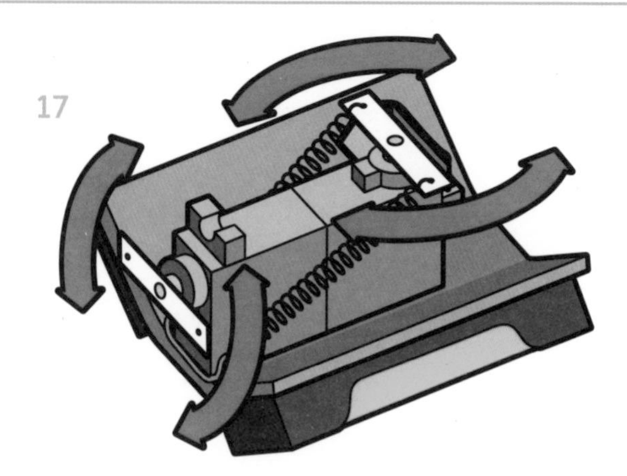

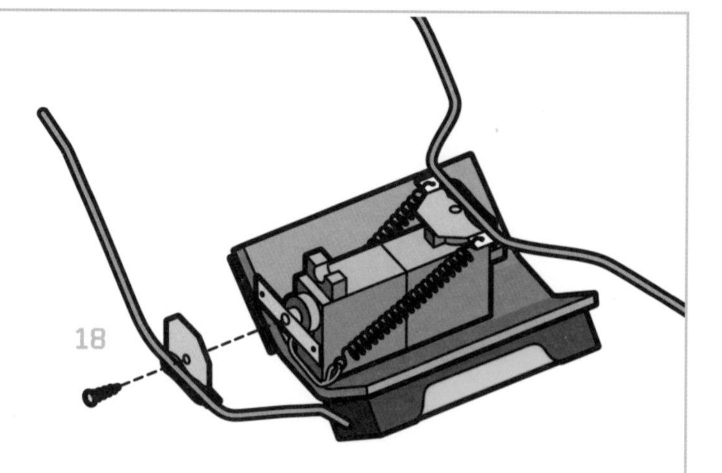

17 Wire the batteries and motors to the circuit board (as shown in the circuit diagram in step 13). When all wiring is complete, the testing can begin. Insert the batteries in the machine and switch ON. It is useful to leave the legs unattached so that the movement of the white nylon actuator arms may be studied without the long brass legs getting in the way. Both actuators should move in back and forth, changing direction every second. The green LEDs should flash alternately as the motor reverses.

18 Finally, screw in the leg assemblies and bend over the brass sheet to lock them in place. The tension in the springs may be adjusted by changing the length of the copper wire, until the rear legs smoothly oscillate about the straight ahead position.

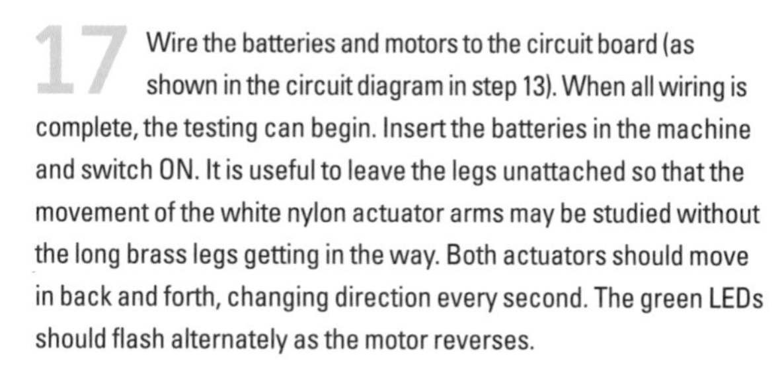

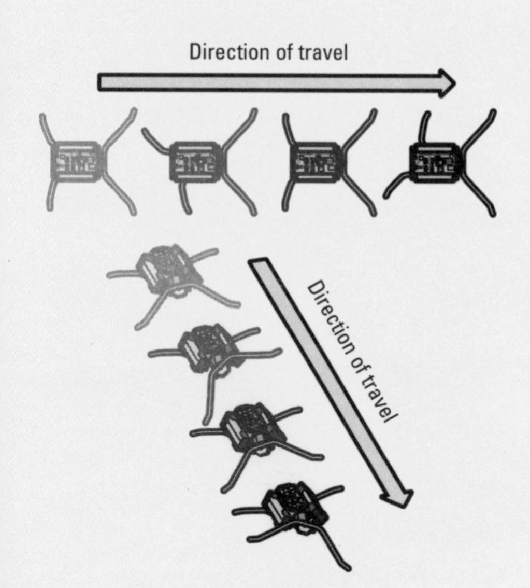

How to Use It

Stomper is now ready to stomp. Find a flat surface, preferably one with a bit of grip, such as a carpet (shiny surfaces will be too slippery for it to perform at its best), and watch your machine shuffle around in the manner of a bad-tempered spider.

Amending the Gait

Stomper is propelled forward by its thin rear legs, and the distinctive stomping gait of its front legs determines where it goes. You can modify its gait by changing over the front wires to the rear motor.

Dealing with Flips

Although the FWD and REV times of the legs are nominally identical, the front legs will sometimes move from the starting horizontal position, due to differences in the motor response in forward and reverse directions, making Stomper flip itself over onto its back. The legs can be restored to the normal horizontal position by turning the machine OFF and gently rotating them.

Carpet Crawling-Caterpillar

There's something hypnotic about the way caterpillars move. They pull the front portion of their bodies forward, then rest while the back half catches up using a ratchetlike motion. Our caterpillar has the same gait. It moves best on a slightly textured surface, such as a carpet, using its tiny metal "claws" to get get a grip. The circuitry that makes it walk is slightly fussy to assemble, but not complex. Take a little time to seek out the perfect carapace. Perhaps use a cut-up milk container decorated in shades of green paint. Alternatively, cover with a thin silicon baking mat. This has an insectlike translucency and a slight rubbery movement—it really does look like larva!

YOU WILL NEED

Medium-density fiberboard (MDF) to make up:

- A: piece, $2\frac{3}{4} \times 1\frac{1}{8} \times \frac{1}{4}$ inches (70 × 28 × 6mm)
- B: piece, $2\frac{3}{4} \times \frac{3}{4} \times \frac{1}{4}$ inches (70 × 22 × 6mm)
- C: piece, $2\frac{3}{4} \times \frac{3}{4} \times \frac{1}{4}$ inches (70 × 22 × 6mm)
- D: piece, $2\frac{3}{4} \times 1\frac{1}{2} \times \frac{1}{4}$ inches (70 × 39 × 6mm)
- E: piece, $1\frac{5}{8} \times \frac{5}{8} \times \frac{1}{2}$ inches (41 × 17 × 12mm)—block for the motor/gearbox
- F: piece, $2\frac{3}{4} \times 5 \times \frac{1}{2}$ inches (70 × 126 × 12mm—for the base
- G: piece, $1\frac{1}{4} \times 1 \times \frac{1}{2}$-inch (32 × 22 × 12mm)
- H: piece, $\frac{3}{8} \times \frac{5}{8} \times \frac{1}{4}$ inch (10 × 20 × 6mm—for the switch

- $\frac{1}{4}$-inch (6-mm) full hex nut
- Two $6\frac{1}{4}$-inch (160-mm) lengths of $\frac{1}{8}$-inch- (2.5-mm) diameter brass or steel rod
- 1-inch (25-mm) length of $\frac{1}{2}$-inch- (12-mm) diameter dowel
- 4-inch (100-mm) length of $\frac{1}{4}$-inch- (6-mm) diameter threaded rod
- Two 1-inch (25-mm) no. 4 wood screws
- Piece of 26-gauge (0.4mm) brass sheet (or metal from an empty can), 2 x 2 inches (50 × 50mm)
- Silicone baking mat, 7 × 7 inches (180 × 180mm)—for the body cover
- Staples

Alternatively (for the body cover):

- 1-quart (1.14l) plastic milk container
- Paint (for decoration, if required)

Electronic components:

- 0.1-inch- (2.54-mm) pitch matrix circuit board, 2 x $1\frac{1}{2}$ inches (50 x 38mm)
- 2K2 $\frac{1}{4}$ W resistor (R1)
- Two 2.2 nF ceramic capacitors (C1, C2)
- 100 nF ceramic capacitor (C3)
- 1N4001 Diode (D1)
- Two 2N7000 MOSFETs (Q1, Q2)
- 5v relay, D.P.D.T. relay
- Toggle switch SW1
- Motor/gearbox, Pololu BCM 60:1 (available at www.technobots.co.uk)

TOOLS

- Adhesive tape
- Craft knife or scissors
- Hacksaw or saber saw
- Hand drill with a good selection of bits
- Industrial-type scissors or tin snips
- Sandpaper and sanding block
- Screwdriver
- Soldering iron (with cored solder and flux)
- Two-tube pack of epoxy adhesive; one tube is the resin, the other is the hardener
- White glue

How to Make the Caterpillar

The illustration on the right illustrates the principle behind the design. It can be seen from 1A (side view) and 1B (plan view) that a motor/gearbox is connected to a leadscrew at one end and a captive nut secured to a front assembly. When the motor is operated, the gap between the two parts is closed, as shown in 2A and 2B, with spikes gripping the carpet and preventing the front assembly from moving. Not visible on this drawing is a simple switch that closes when the rear and front parts touch. This contact, through a simple piece of circuitry (shown later), causes the motor to reverse. Because the rear assembly is also fitted with spikes, it cannot move backward, and, as a result, the front assembly is pushed forward. When it reaches the end of its travels shown in 1A, another switch is closed, which again reverses the motor. In this way the machine slowly pulls itself forward.

1 The wiring of the circuit board is straightforward. The operation is very simple, because it is a bistable circuit where either Q1 or Q2 is ON, thus opening and closing the 5v relay, the function of which is to reverse the motor. The FWD and REV switches operate the relay at the end of the travel of the main frame.

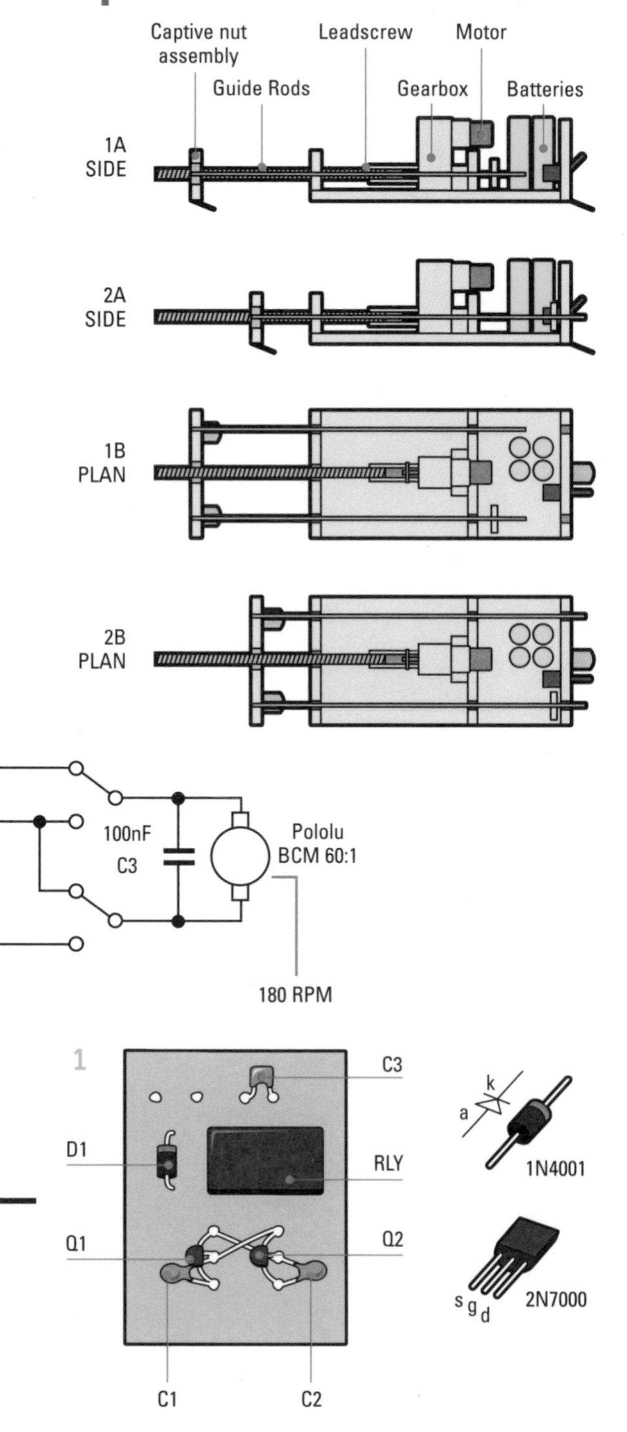

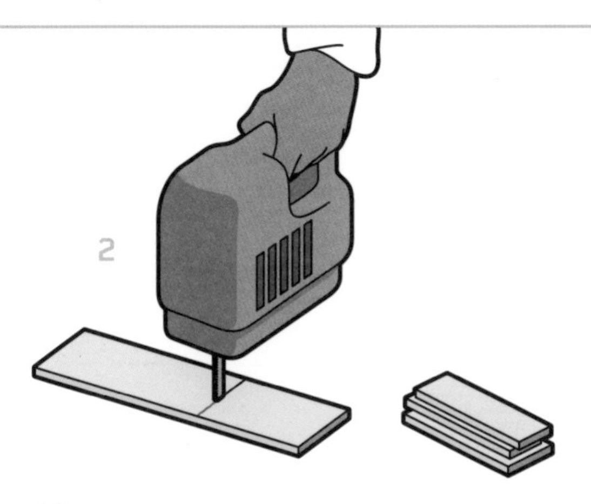

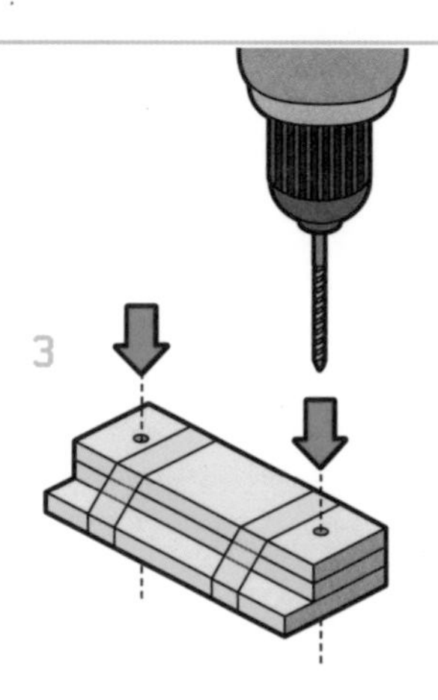

2 Cut pieces A, B, C and D from the ¼-inch (6-mm) MDF. Use adhesive tape to fasten A, B, and C together with the long edges lined up.

3 Drill holes, as shown (drilling must be performed as a single operation to ensure that the spacing between the holes is the same on all pieces; this ensures in turn that the guide rods will slide smoothly over the length of travel).

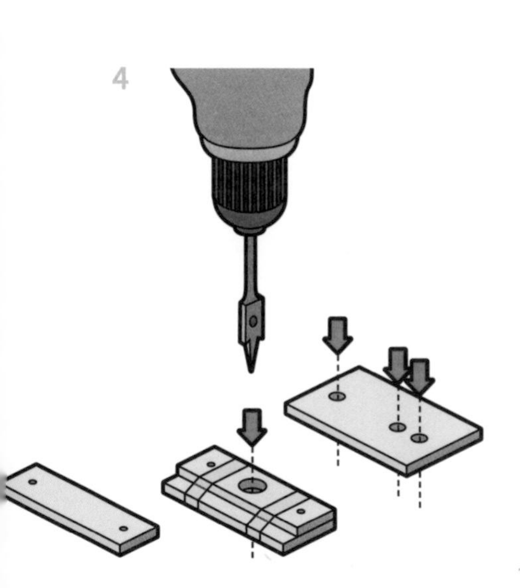

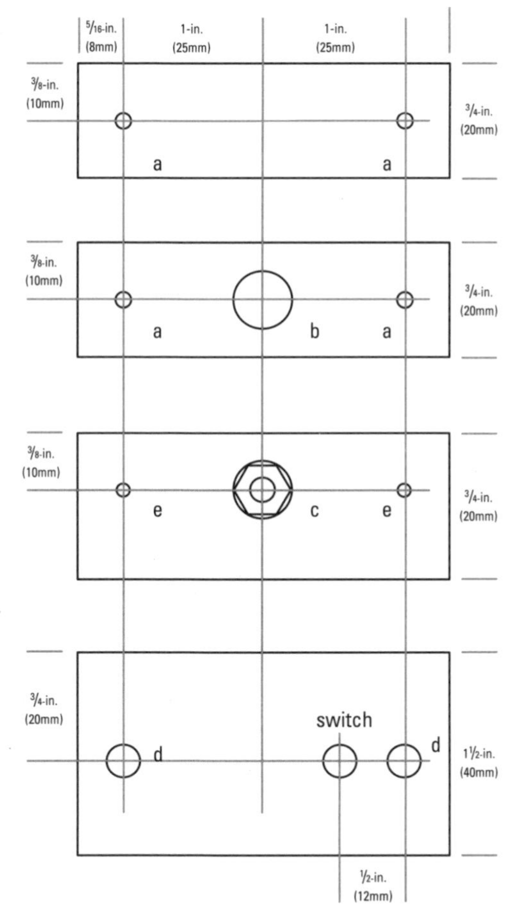

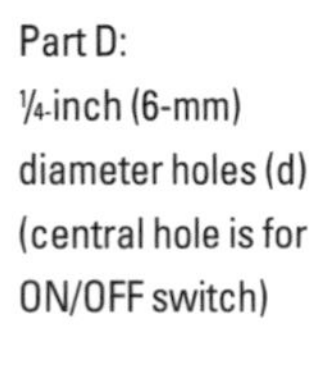

Part C:
⅛-inch (3mm) diameter holes (a)

Part B:
Small holes (a)
⅛-inch (3mm)
Big hole (b)
⅜-inch (10mm)

Part A:
Small holes (e)
⅛-inch (3mm)
Big hole (c)
⅜-inch (10mm), file to adjust and epoxy ¼-inch (6-mm) nut

Part D:
¼-inch (6-mm) diameter holes (d) (central hole is for ON/OFF switch)

4 Remove piece C; then, keeping A and B together, drill a ½-inch (12-mm) hole, centered, as shown. Drill clearance holes for the guide rods in piece D (precision here is not so critical) and another ¼-inch (6-mm) hole to mount the ON /OFF switch.

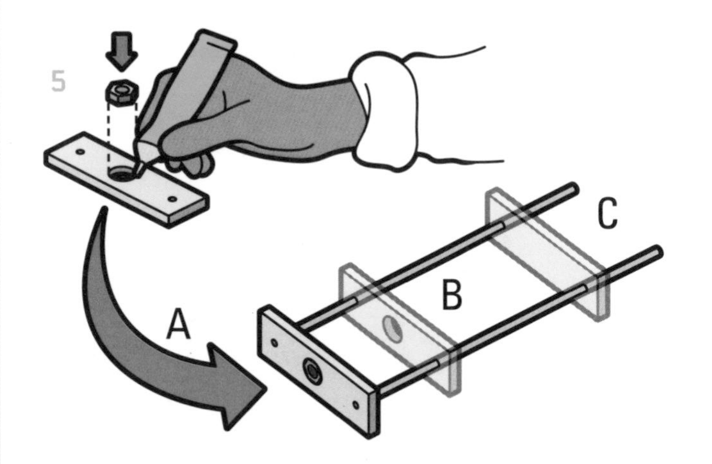

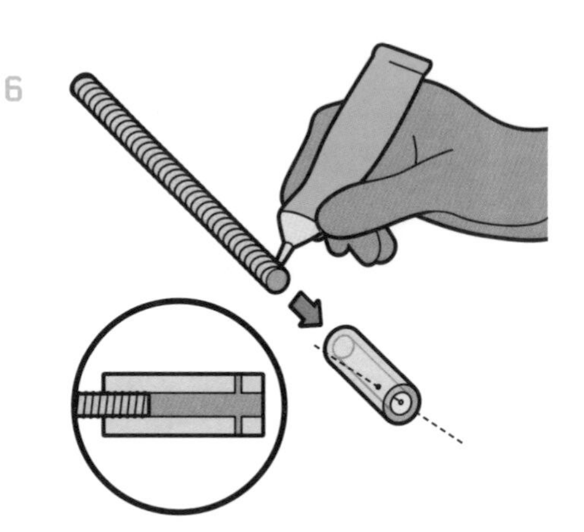

5 Carefully file the ½-inch (12-mm) hole in piece A until an ¼-inch (6-mm) nut can be fitted through it; then epoxy the nut in place. Epoxy the brass rods into the holes in piece A and locate the other ends in pieces B and C. This ensures that the rods will be set at the correct spacing. Let the epoxy harden.

6 Drill a ¼-inch (6-mm) hole down the center of the 1-inch (25-mm) piece of dowel (this functions as the coupler between the gearbox and the threaded rod). Insert the threaded rod about ⅜ inch (10mm) into the dowel and epoxy in place. Let it harden. When hardened, make a 1/16-inch (2-mm) hole in the coupler, as shown.

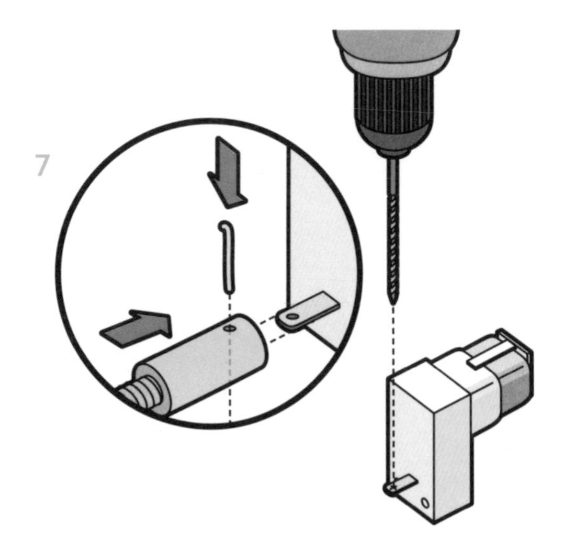

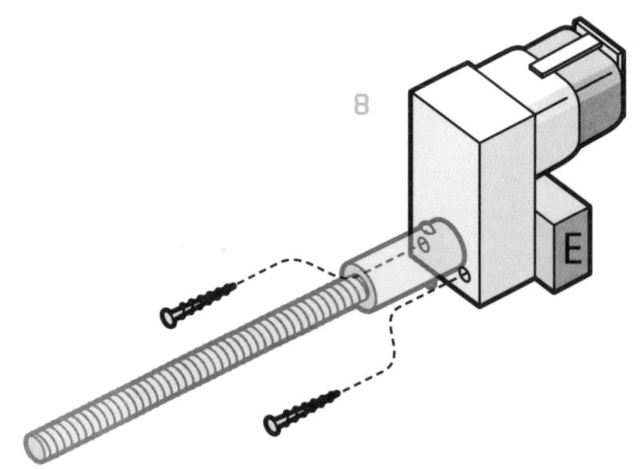

7 Drill a 1/16-inch (2-mm) hole in the gearbox output shaft as shown above. Insert, a piece of steel wire through the coupler and through the hole in the shaft, this then transmits the drive from the motor/gearbox into the threaded rod (see next step).

8 Attach piece E to the motor/gearbox with two 1-inch (25-mm) screws. Pass the screws through the holes just below the gearbox output shaft.

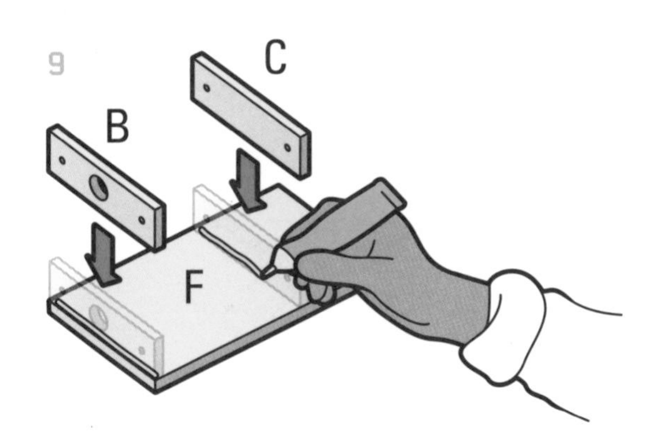

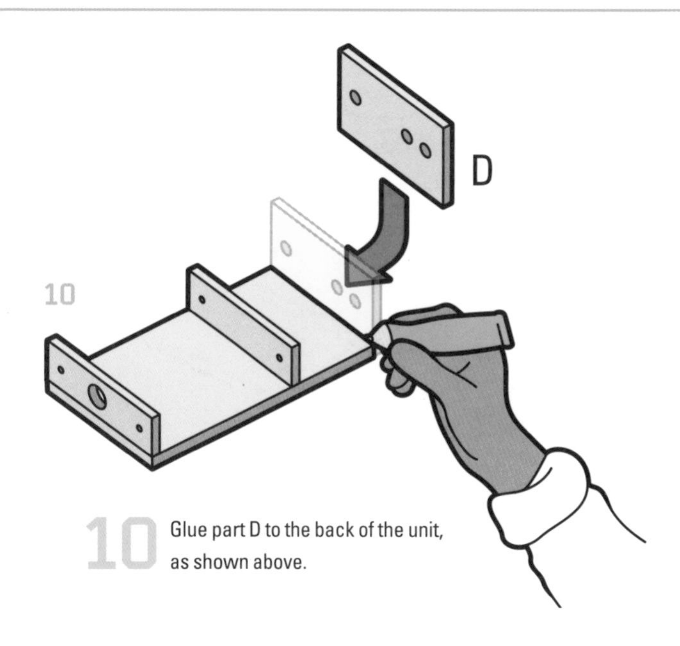

9 Using white wood glue, secure parts B and C in place on base piece F, allowing a space of 3 inches (75mm) between them (piece A remains secured only to the brass rods, allowing it to move backward and forward).

10 Glue part D to the back of the unit, as shown above.

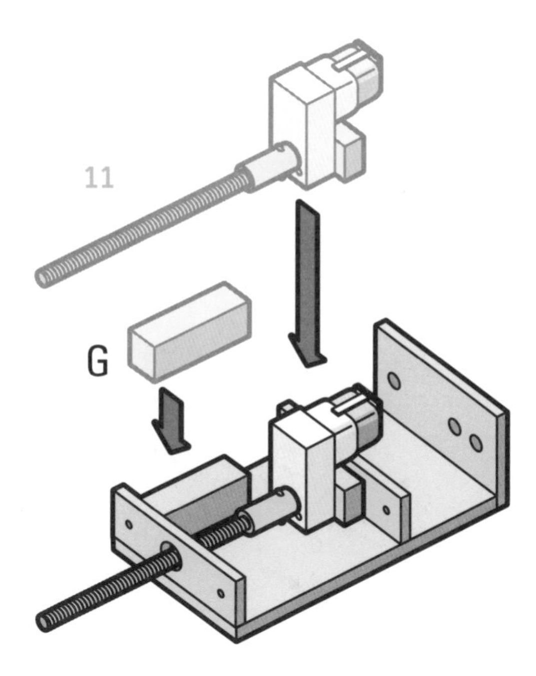

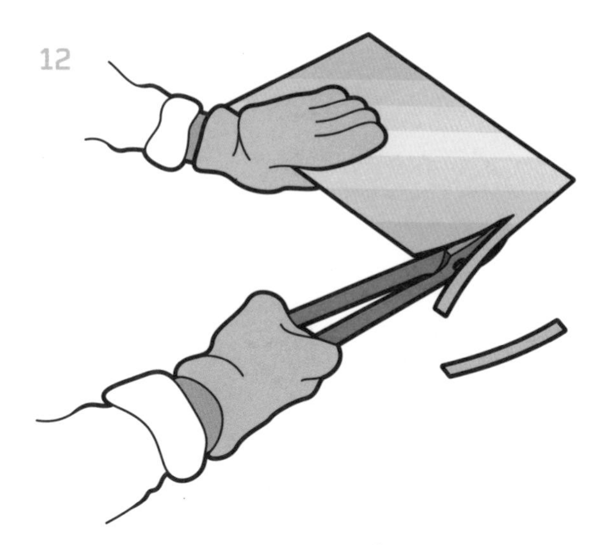

11 With the motor and gearbox in place, glue piece G in place as shown.

12 To make the switches, use the tin snips to cut strips approximately ¼ inch (6mm) wide from the brass sheet (or metal from a can).

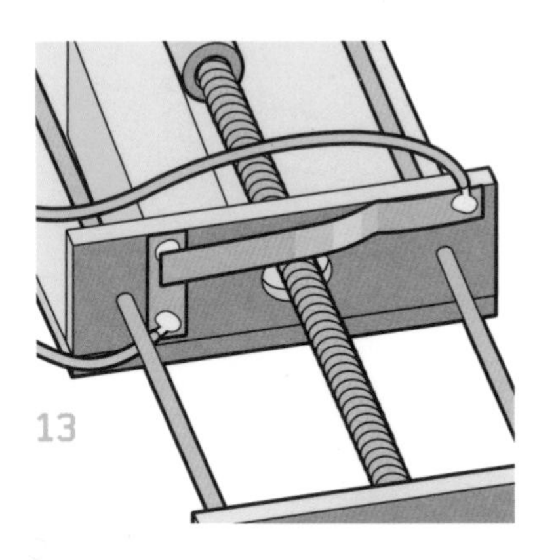

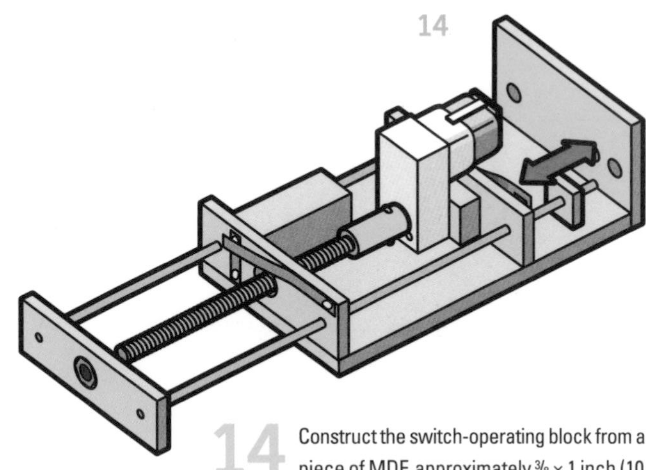

13 Two parts are needed for each switch: the fixed contact, shown here connected to the left-hand wire, and the long moving contact, at the top of the block. Apply blobs of solder to the contact points (to avoid oxidization of the brass, which would result in a poor contact). Exact dimensions are not critical (only the setting) and the metal should just be bent to the general form as shown here.

14 Construct the switch-operating block from a ¼-inch (6-mm) piece of MDF, approximately ⅜ × 1 inch (10 × 25mm), as shown, with a $^1/_{16}$-inch (2.5-mm) hole drilled in it to make it a tight fit on the slide rods (when this block contacts the rear switch, the motor reverses). You can adjust the position of this to make it a long-stroke or a short-stroke machine; the nearer it is pushed toward the front the shorter the stroke, and vice versa. The piece of MDF moves in the direction of the arrows.

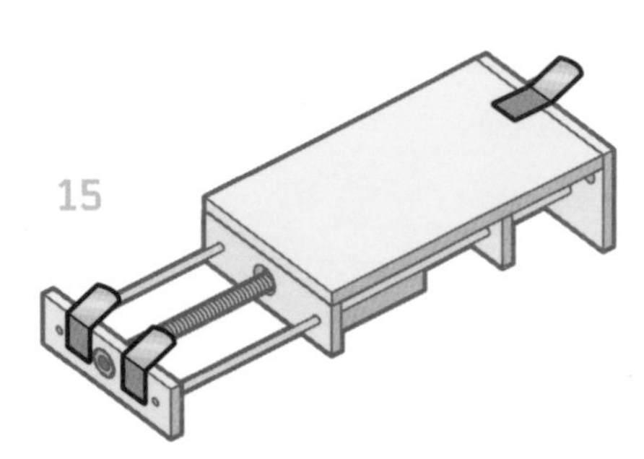

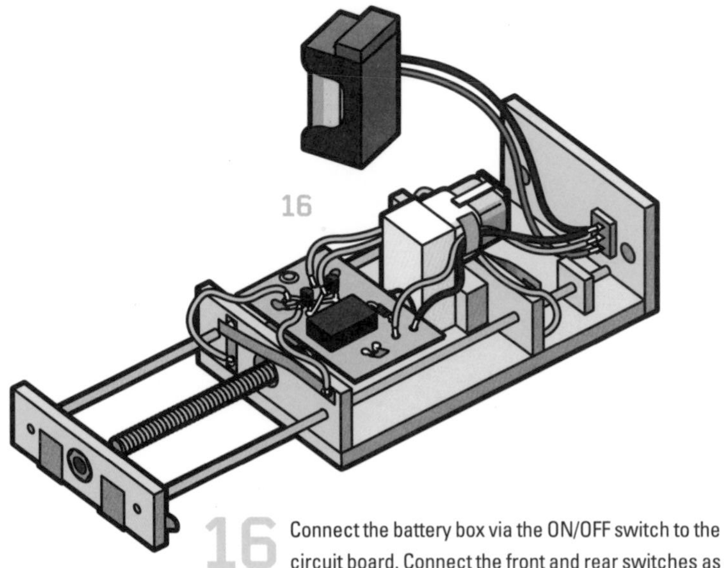

15 To make the claws needed for the machine to grip the carpet, cut three small lengths approximately 1½ inches (40mm) long (two for the front and one for the back) from same brass sheet used for the switches. Attach with glue to the front and back panels. They should be angled slightly downward, for maximum grip required to move on the carpet.

16 Connect the battery box via the ON/OFF switch to the circuit board. Connect the front and rear switches as shown on the circuit diagram. The front switch is wired from the contacts visible on the following steps.

17 Take the wires from the reversing relay and touch them onto the motor terminals. Make the motor rotate so that it will engage the threaded rod into the captive nut on part A. The thread should be engaged for about ¾ inch (20mm).

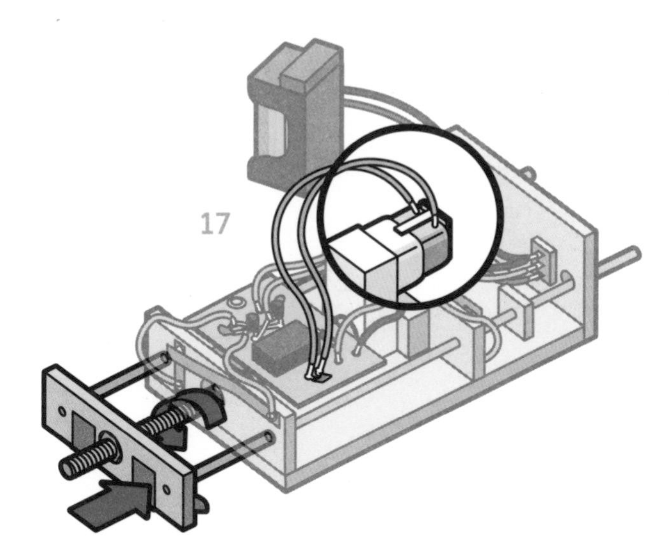

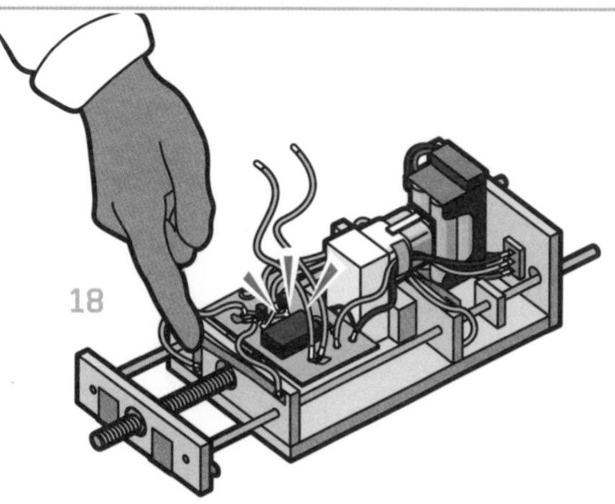

18 Disconnect the wires to the motor and alternately press the front and rear switches, making certain that they have a gap of about 1/32 inch (1mm) before you start. The relay should give a click as each alternate switch is pressed, confirming the circuit is working and that the relay will reverse the power to the motor when connected.

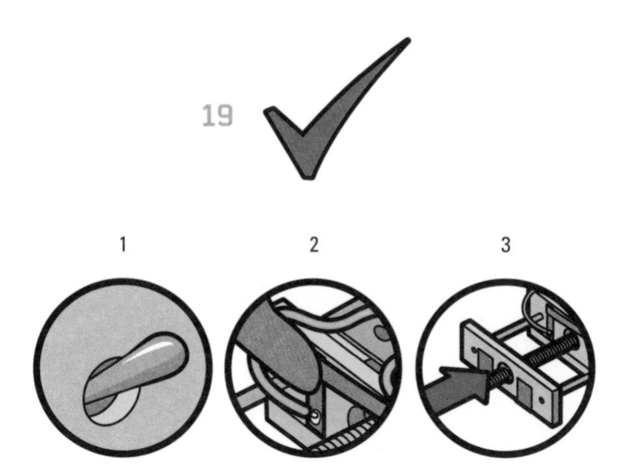

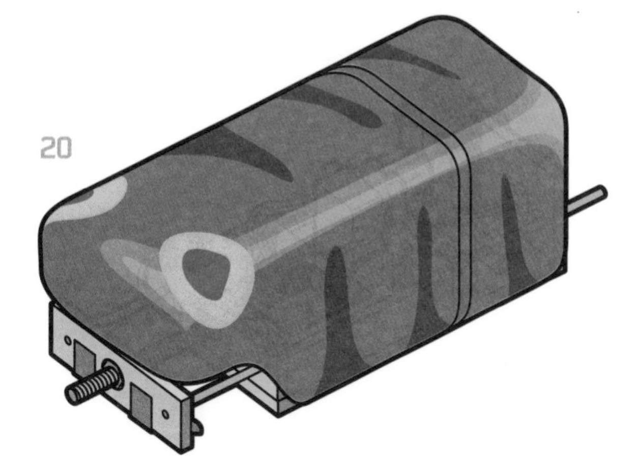

19 Switch OFF and connect the wires to the motor. Switch ON and the motor will start to move part A with respect to the main body. Quickly operate the front switch and if the motor the starts to close the part A to the main body, then the motor is wired correctly. If to achieve this reversal you need to operate the rear switch, then simply change over the wires to the motor.

20 To make the caterpillar body, a simple solution is to cover the caterpillar with thin, silicone baking material (see page 107), forming the curve of the caterpillar and stapled to the base. But if you are feeling more creative, carefully use a craft knife or pair of scissors to trim a standard 1-quart (1.14l) plastic milk container to size to fit over the structure. Finally, paint or decorate to your own design as above.

How to Use It

After building your carpet caterpillar, do not forget the most important point—that it is designed to crawl over carpets. A slippery surface, such as tiles, will cause a problem because they do not provide anything for the claws to dig into.

Troubleshooting

The electronics controlling the caterpillar are simple, but the direction of rotation of the motor is most important, and steps 18 and 19 are the most critical ones. If after a time the caterpillar fails to operate correctly, clean the switch contacts by pressing then together with a piece of clean paper sandwiched in-between. Pull out the paper and your contacts will be clean.

Carpet-Crawling Caterpillar

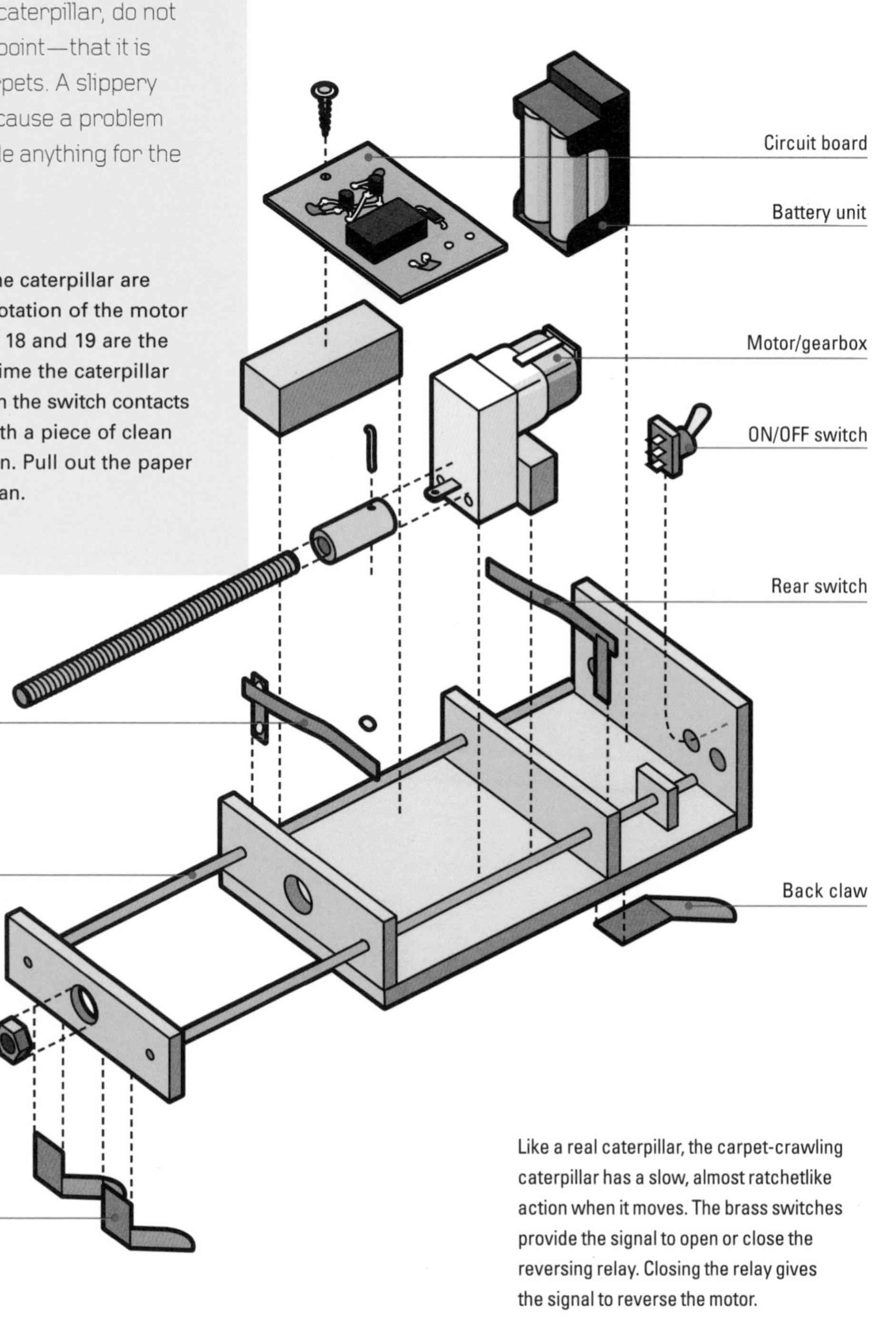

Like a real caterpillar, the carpet-crawling caterpillar has a slow, almost ratchetlike action when it moves. The brass switches provide the signal to open or close the reversing relay. Closing the relay gives the signal to reverse the motor.

MACHINES THAT TALK

Electronic Didgeridoo

The true aboriginal didgeridoo, made from hollowed wood, produces a haunting, droning music. We wanted to make its electronic equivalent, and we managed to create a remarkably similar noise—by causing a pipe to resonate acoustically. Even better, this wind instrument doesn't need great lung capacity or skill; it's operated by turning a switch on and off, although a lot of the variation of the harmonies you can achieve will depend on how good you get at "tuning" it to the best frequency for playing. The circuitry is made on a matrix board, which simplifies the wiring and helps to avoid accidental connections. Allow a few hours to make your didgeridoo, and two more to learn to play it well enough to delight your friends.

YOU WILL NEED

MDF to make up:

- A: piece, 4 × 4 × ½ inches (100 × 100 × 12mm)
- B: piece, 5½ × 3½ × ½ inches (140 × 90 × 12mm)
- C: piece, 3½ × 1⅜ × ½ inches (90 × 35 × 12mm)
- D: piece, 3½ × 1⅜ × ½ inches (90 × 35 × 12mm)
- E: piece, 4½ × 1⅜× ¼inches (115 × 35 × 6mm)
- F: piece, 4½ × 1⅜ × ¼ inches (115 × 35 × 6mm)
- G: piece, 5½ × 3½ x ½ inches (140 × 90 × 12mm)
- H: piece, 1 × ½ × ½ inches (25 × 12 × 12mm)
- I: piece, 1 × ½ × ½ inches (25 × 12 × 12mm)
- J: piece, 3 × ½ × ½ inches (80 × 12 × 12mm)
- K: piece, 3 × ½ × ½ inches (80 × 12 × 12mm)
- 39½-inch to 10-foot/1,000- to 3,000-mm length of 2¾-inch (69-mm) PVC standard drainpipe tube in black
- Black paint and brush
- Four 1-inch (25-mm) screws, ⅛ inch (4mm) in diameter, and 12 nuts (for the speaker)
- Two 1½-inch (40-mm) screws, ⅛ inch (4mm) in diameter, cut to ¾-inch (20-mm) lengths with the heads removed, and two nuts (for the lid)
- Four ¾-inch (20-mm) no. 4 wood screws (to attach two to each to pipe and circuit board)

Electronic components:

- 0.1-inch- (2.54-mm) pitch matrix circuit board, 3¼ x 2½ inches (81 x 63mm)
- 47K ¼ W resistor (R1, R2)
- 22K ¼ W resistor (R3)
- 470K ¼ W resistor (R4)
- 220K ¼ W resistor (R5)
- 1M horizontal preset (RV1 RV2)
- 10nF polyester capacitor (C1)
- 2.2uF 63 v capacitor (C2)
- LM 358N, dual op. amp
- IRFZ24N MOSFET (Q10)
- 3½-inch- (90-mm) square bass loudspeaker
- Battery holder (for four AA batteries)
- PP3 connector (for above)
- Main ON/OFF switch: single pole, single throw (SW1)
- Momentary push button drone switch (SW2)

TOOLS

- Compass (geometrical)
- Cordless drill with a good selection of drill bits to fit
- Epoxy adhesive
- File
- Hacksaw
- Hand saw
- Pencil and ruler
- Pliers
- Saber saw
- Side cutters
- Soldering iron (with cored solder and flux)
- Tenon saw
- Wrenches

How to Make the Electronic Didgeridoo

1 Using the hacksaw, cut a length of the PVC tube and sand down any rough edges. For a rich, deep note, you'll need at least 39½ inches (1,000mm), but lengths up to 10 feet (3,000mm) can be used; however, the longer the instrument, the harder it is to carry.

2 Cut a 4-inch- (100-mm) square block of ½-inch (12-mm) MDF using either the saber saw or handsaw as piece A. Pencil lines from corner to corner to find the center, then spike the compass in the center, set at a radius of 1⅜ inches (35mm) and scribe a 2¾-inch (70-mm) circle. Use the saber saw to cut a hole in the center. File smooth the inside of the hole, checking it is wide enough to accept the length of PVC tube. If not, a little more filing may be needed.

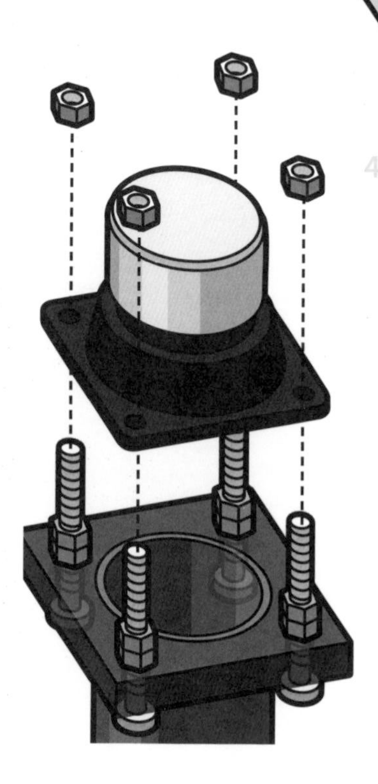

3 Position the loudspeaker centered on the block, pencil around this, and then mark on the block the position of the mounting holes on the loudspeaker. Drill ¼-inch (6-mm) holes through these.

4 Paint the base black to match the tube, if desired. Epoxy the block and tube together so that the end of the tube fits flush against the edge of the block. Using four ⅛-inch (4-mm) bolts and twelve nuts, attach the loudspeaker to the MDF block, with a gap of approximately ⅜ inch (10mm) between the two. Tighten up the nuts with a wrench.

5 Using a tenon saw or saber saw, cut the following pieces of ½-inch (12-mm) MDF: piece B (for the base of the circuit board housing), 5½ × 3½ inches (140 × 90mm); and pieces C and D (for the short end pieces), 3½ × 1⅜ inches (90 × 35mm). Use the sanding block to smooth all edges. Cut the following pieces of ¼-inch (6-mm) MDF: pieces E and F (for the long sides of the box), 4½ × 1⅜ inches (115 × 35mm); and piece G (for the lid), 5½ × 3½ inches (140 × 90mm). Again, smooth all the edges. For mounting the circuit board, cut two small blocks of ½-inch (12-mm) MDF, each 1 × ½ inch (25 × 12mm), which are pieces H and I.

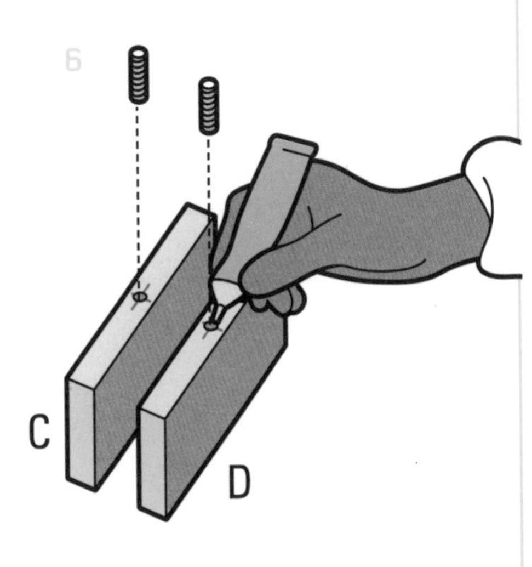

6 Place pieces C and D on their narrow edge, measure and mark the center from end to end, and then at each point drill two ⅛-inch (4-mm) holes about ⅜ inch (10mm) deep. Use two ⅛-inch (4-mm) studs (cut to length) and epoxy to fasten these in the holes.

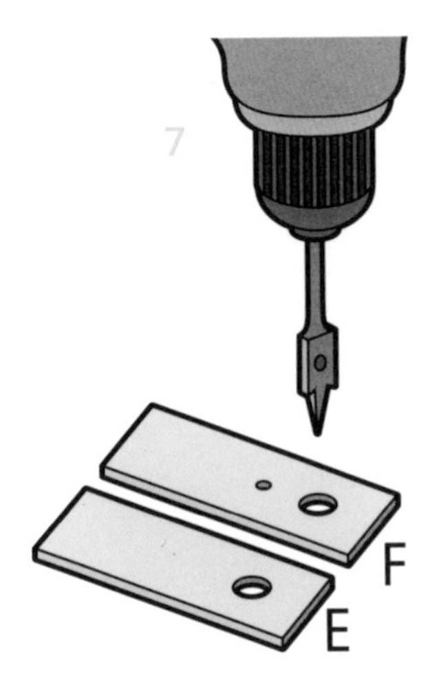

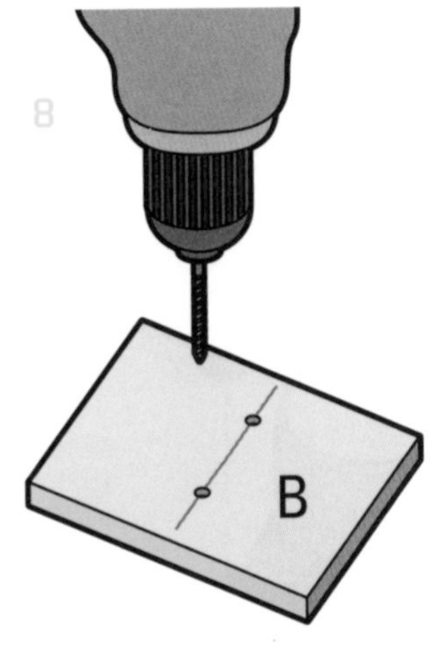

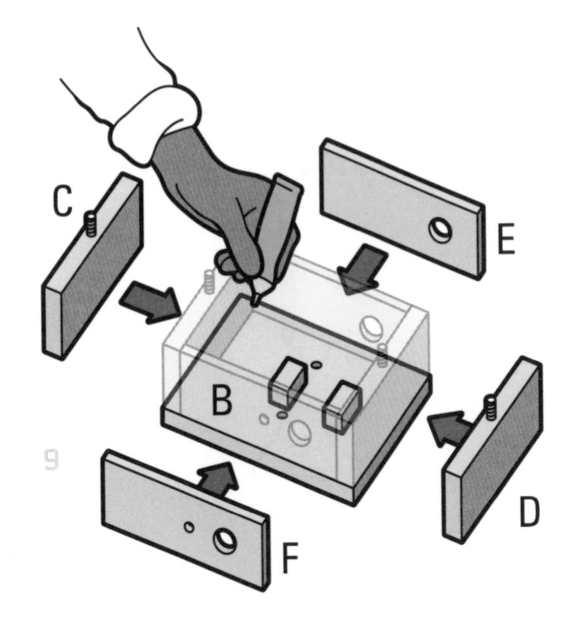

7 Drill holes in pieces E and F to accommodate the switches (the exact size and location of the holes will depend on the particular switch employed. Space must be left to fit the circuit board), and on piece F drill a ¼-inch (6-mm) hole to the left of this for the speaker cable.

8 Measure the center of piece B, lengthwise, and draw a centerline. Measure two points on this and drill two 1-inch (25-mm) holes.

9 Glue pieces C, D, E, and F onto piece B, then glue the mounting blocks to hold the circuit board, leaving enough space for the battery holder next to piece C to slot in.

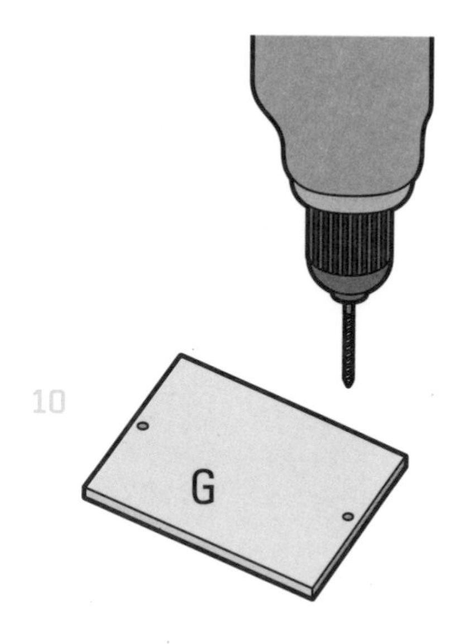

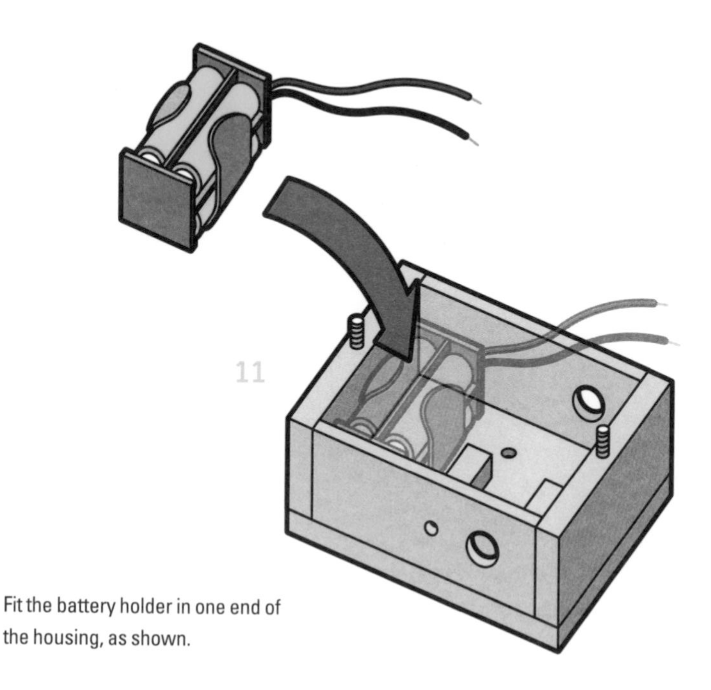

10 Drill two ¼-inch (6-mm) holes in the lid (piece G) to line up with the two ⅛-inch (4-mm) studs.

11 Fit the battery holder in one end of the housing, as shown.

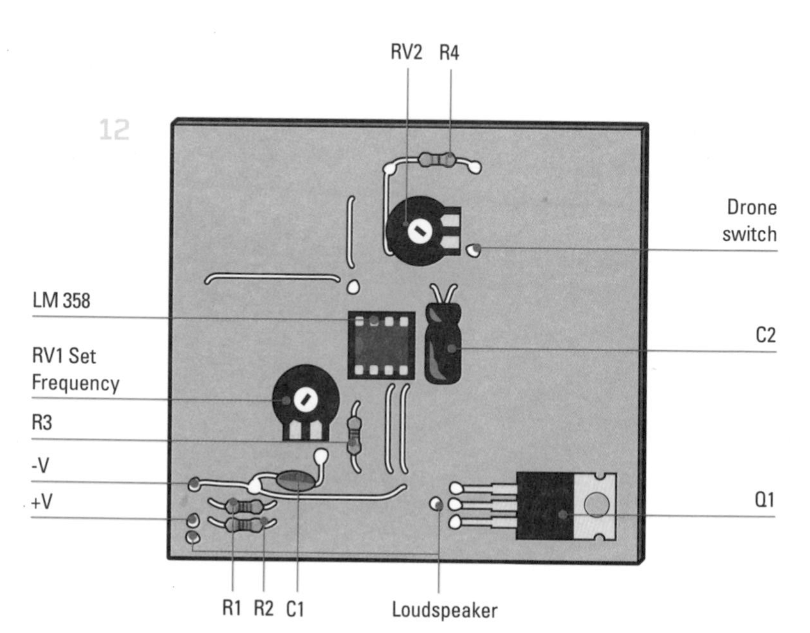

The base connections for these parts are shown below:

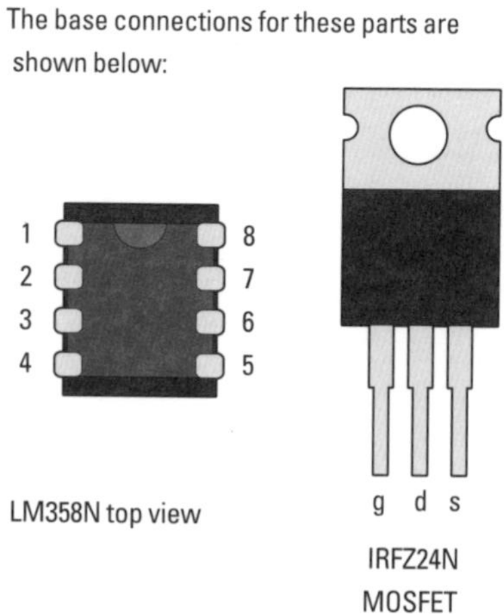

12 Assemble components on the circuit board, as shown (the exact layout of the components is not critical but it is suggested that the illustration is generally followed because it will keep the wiring simple). You will need the pliers and side cutters to help you put together the components.

Please note: LM358N and the IRFZ24N MOSFET are both static sensitive, so before handling them, wrap a length of wire around your wrist and connect this to a known ground point, such as a faucet.

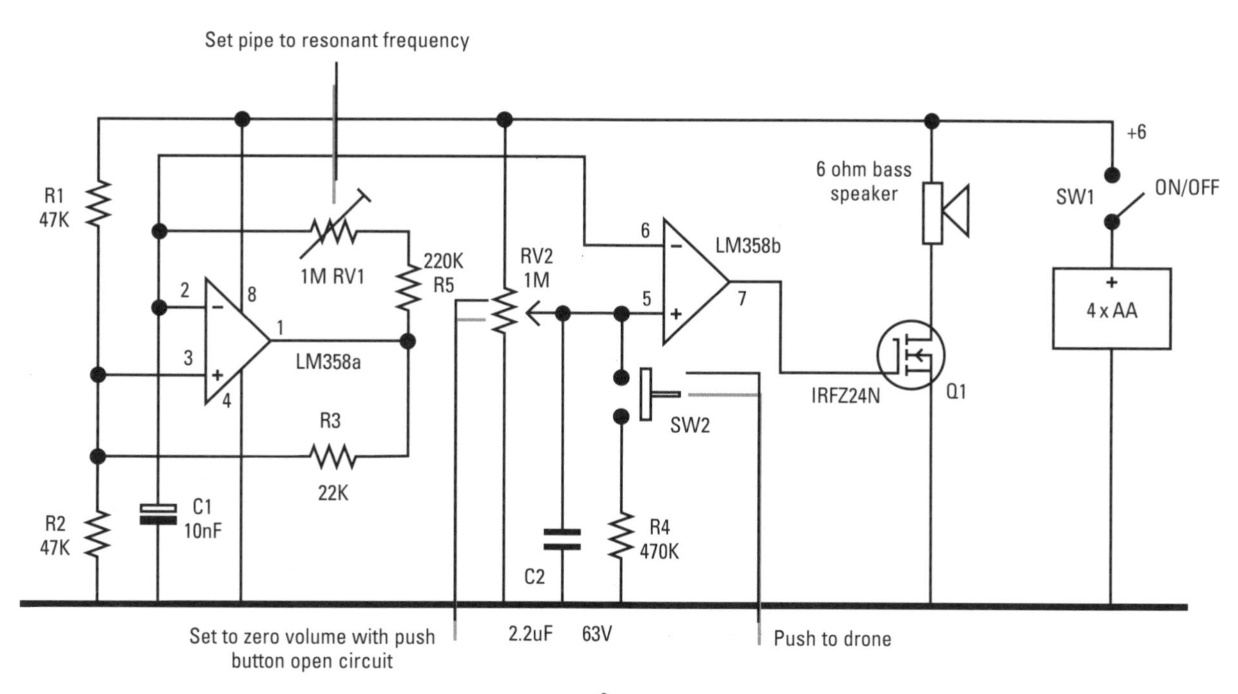

13 Fit the ON/OFF and drone switches (following the manufacturer's instructions for the model you've chosen).

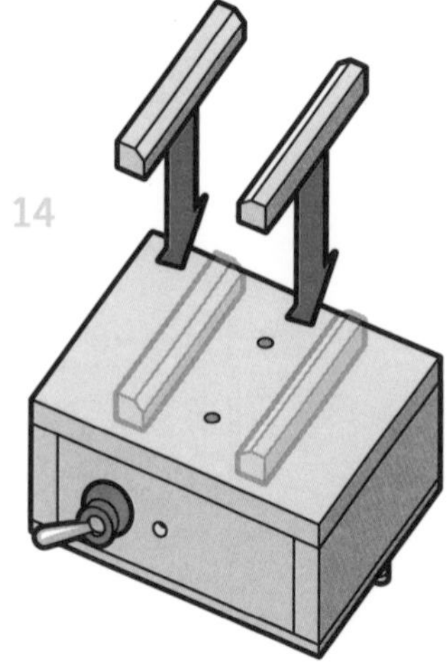

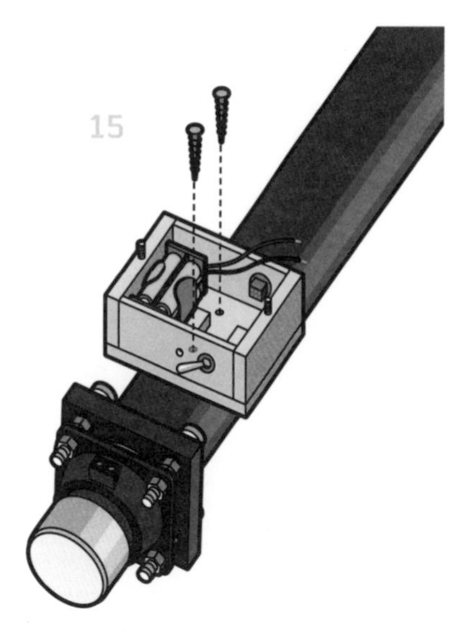

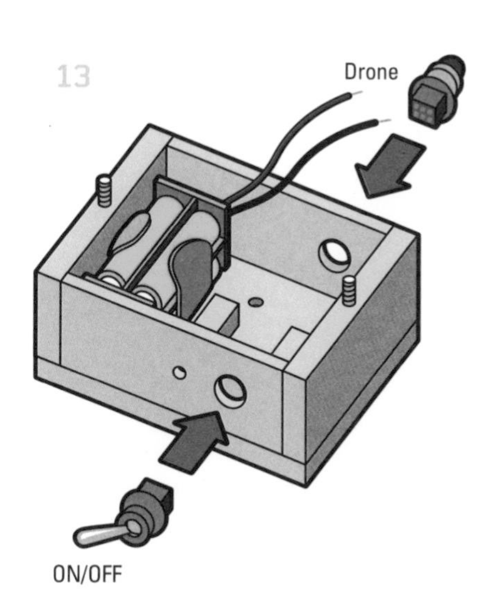

14 Add pieces J and K in ½-inch (12-mm) MDF, 3 x ½ inches (80 x 12mm). Glue to the bottom of the box with an 2-inch (50-mm gap) in between. This wood is to steady the pipe and make it more secure when it is attached. You may need to remove the inner edges of the wood in order for the pipe to sit snugly in between.

15 Drill two ⅛-inch (4-mm) holes along the centerline of the bottom of the box and then screw this to the pipe using two ¾-inch (20-mm) no. 4 wood screws.

The Electronic Didgeridoo

16

16 Drill two ⅛-inch (4-mm) holes in the mounting blocks at the bottom of the box. Screw the circuit board to the wood using two ¾-inch (20-mm) no. 4 wood screws.

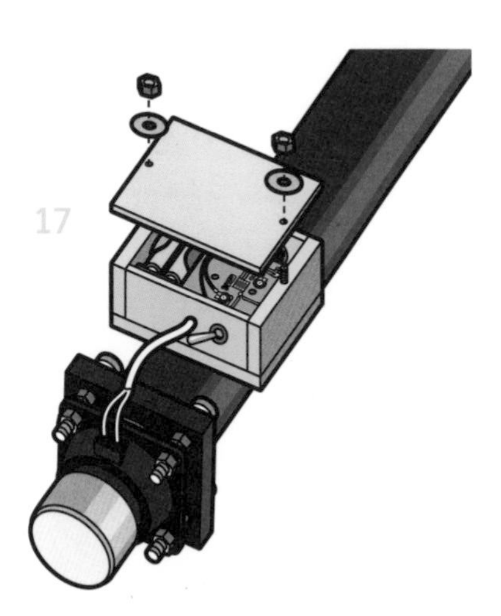

17

17 Connect the loudspeaker wire to the circuit board, running it through the hole drilled in piece F. Secure the lid on the studs using two ⅛-inch (4-mm) nuts, tightening them with wrenches.

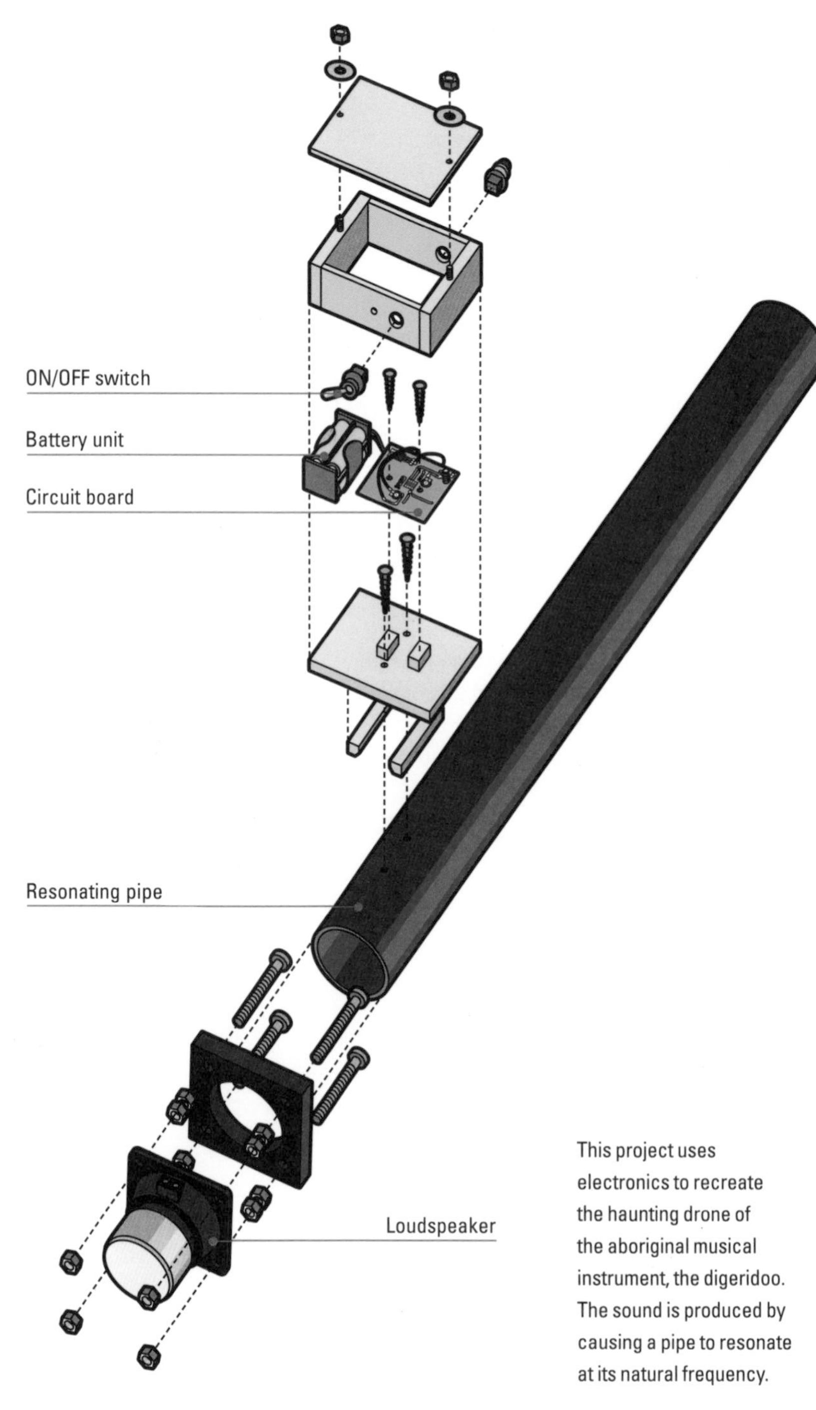

This project uses electronics to recreate the haunting drone of the aboriginal musical instrument, the digeridoo. The sound is produced by causing a pipe to resonate at its natural frequency.

How to Use It

It now simply remains to tune the pipe to the correct frequency. The basic frequency is produced in the op.amp LM358a, the frequency being set by RV1. A range of approximately 10Hz to 1KHz is possible and you may "tune" the instrument to the length of pipe in use.

Adjusting the Sound

A triangular waveform is present at pin 2, with an amplitude of 3 volts maximum and 1.5 volts minimum. This is the linked to a second op. amp. LM358b, which is used as a comparator. RV2 is adjusted to set to a nominal 3 volts at pin 5 and in this setting the triangular waveform applied to pin 6, does not reach the switching threshold of the comparator, whose output is coupled to Q1, an IRFZ24N MOSFET which drives the loudspeaker. Adjust RV2 so that it is just on the edge of the speaker producing a sound. When SW2 is operated the voltage at pin 5 will fall to about 2 volts (thus allowing the triangular waveform through) and hence operate the loudspeaker. It will not perform this operation instantaneously, but, due to the action of C2, it will occur slowly over a one- to two-second period, thus imitating the slow rise and fall of the true didgeridoo.

Using the Drone Switch

The sound generated by the speaker is very rich in harmonics and this enables the pipe to be set to more than one possible operating mode and tuning may be left to the user. The operation requires RV1 to be adjusted and, as the frequency is changed, the resonant peaks will be clearly heard because the sound intensity will increase greatly. When set to the chosen drone, operation of SW2, the drone switch, can produce different effects. Pressing it and holding it ON will produce a continuous drone. However, by varying the rate and duration of the ON/OFF cycle, a more modulated note, similar to a true wooden instrument, is produced.

The voicebox of the didgeridoo.

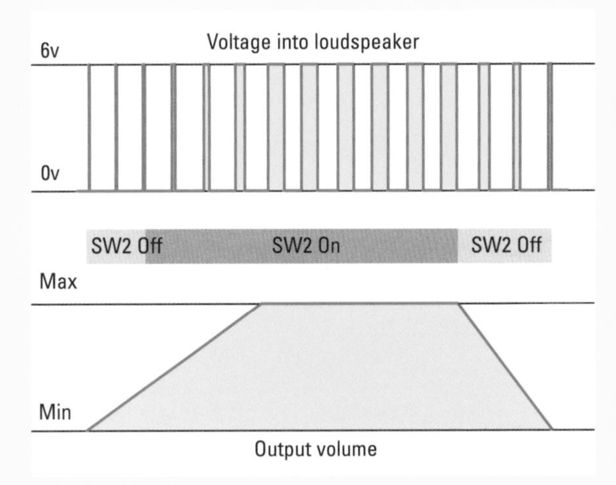

Diagram to show the waveform.

The brains of the didgeridoo.

Drum-Beating Whirligig

The whirligig is really a wind-powered drum, and it's a pleasure to watch it in action. The wind drives the sails, which turn the driveshaft, which prompts the cam trips to press down on the drumsticks and... rat-a-tat-tat! If you're an experienced woodworker and the basic model doesn't look challenging enough for your attention, try widening the box and increasing the number of cam trips, setting them in order around the shaft; with some extra calculation and experimentation, you will find that it's possible to create a whirligig that plays drum rolls. The veneer plywood and close-grained softwood add to the enjoyment of the construction, as both cut to crisp, clean edges.

YOU WILL NEED

Best-quality exterior-grade multicore plywood to make up:

- A: piece, 10 × 28 × 3/8 inches (255 × 710 × 10mm)
- B: piece, 4 3/4 × 7 1/4 × 3/8 inches (120 × 185 × 10mm)
- C: piece, 2 × 7 1/4 × 3/8 inches (50 × 185 × 10mm)
- D: piece, 2 × 7 1/4 × 3/8 inches (50 × 185 × 10mm)
- E: piece, 10 3/4 × 7 1/4 × 3/8 inches (275 × 185 × 10mm)
- F: piece, 10 × 6 3/8 × 3/8 inches (255 × 160 × 10mm)
- G: piece, 8 × 7 1/2 × 3/8 inches (205 × 190 × 10mm)
- H: piece, 6 1/2 × 6 × 3/8 inches (165 × 150 × 10mm)
- Piece of straight-grain pine, 24 × 2 × 3/4 inches (610 x 50 x 20mm) (for the propeller boss)
- 48-inch (1,220-mm) length of 1-inch- (25-mm) diameter dowel (for the driveshaft and pivot pole)
- 18-inch (455-mm) length of 1/2-inch- (12-mm) diameter dowel (for the two drumsticks and various pegs, also allowing generous cutting waste)
- Two strips of lead, about 1 × 1/2 inches (25 × 12mm)
- A good supply of brads and screws
- Four 2-inch- (50-mm) bolts, 3/8 inch (10mm) in diameter, with washers and nuts to fit
- 10-inch (255-mm) length of straightened coat hanger wire
- Eleven 3/8-inch- (10-mm) diameter wooden beads, or washers/buttons/tube (for spacing the sticks on the wire)
- Danish oil
- Scaffold pole to mount the unit

TOOLS

- Adjustable wrench
- Clamps
- Cordless drill/driver with a good selection of drill bits to fit, including a 1 1/8-inch- (28.6-mm) diameter Forstner drill bit
- Craft knife
- Dividers
- Pencil and ruler
- Pliers
- Sandpaper and sanding block
- Scroll saw with plenty of spare blades to fit
- Set of mortise chisels
- Shoulder plane
- Small hammer
- Square
- Straight handsaw
- Workbench with vise

How to Make the Whirligig

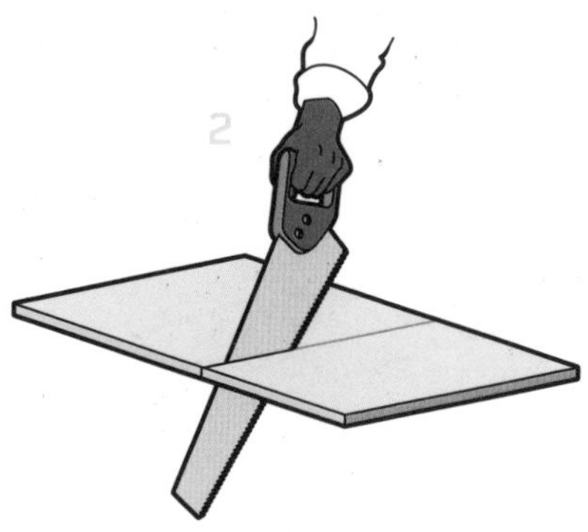

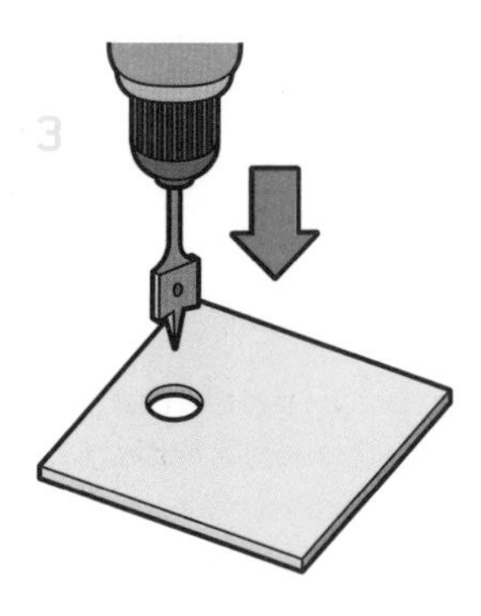

1 Use the pencil, ruler, dividers, and square to draw the L-shaped sides needed for the box onto the plywood. Measurements and shapes for these are provided on page 158.

2 Use the straight handsaw to cut out piece A (the tail vane), piece B (the drum surface), and pieces C and D (the two front frame pieces). These will form part of the primary box form. Sand the sawn edges to a crisp finish.

3 Use the straight handsaw to cut out the following pieces: piece E (the outer base slab); piece F (the inner base slab); piece G (the top slab) and piece H (the inner top slab). Together, these will make up the rest of the primary box form. Use the drill and a 1⅛-inch-(29-mm)- diameter Forstner bit to drill holes in all the box pieces as shown in the templates section on pages 158—159. Again, sand any rough edges.

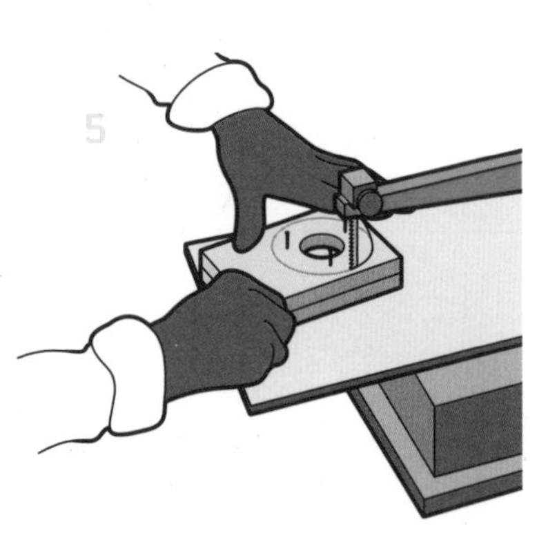

4 Use the straight saw and scroll saw to cut out the four propeller blades. The measurements and shape for these are provided on page 158. Identical multiples can be achieved by stacking and nailing the plywood with brads, so that the stack of blades can be cut altogether). Try throughout to run the line of cut to the waste side of the drawn line.

5 Use the scroll saw and 1⅛-inch- (29-mm) diameter Forstner bit to cut four washers, the large base support disk, and pegs A and B that are used when placing the whirligig in the pivot pole in step 19. Measurements and shapes for these are provided on page 157.

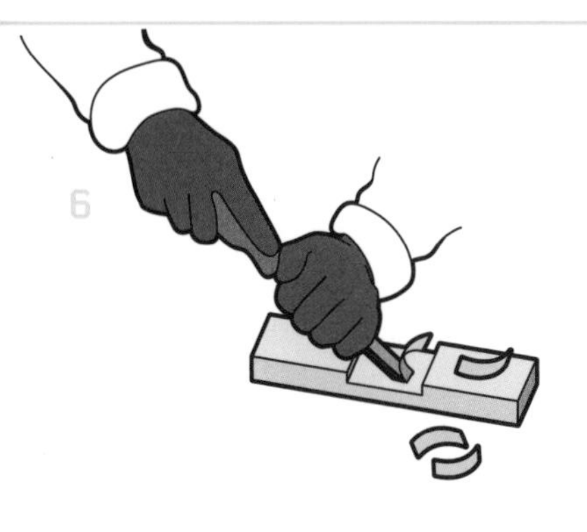

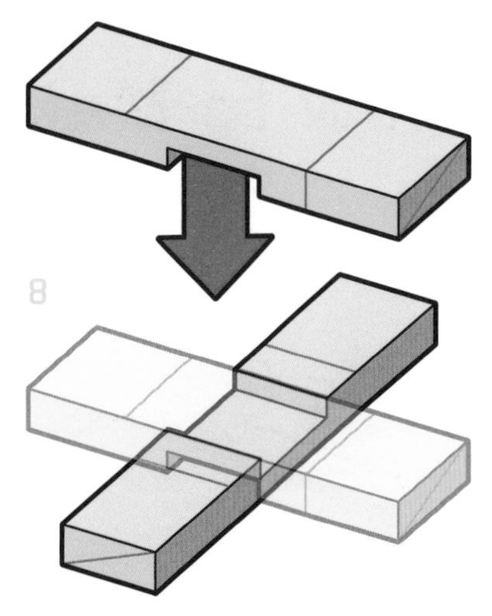

6 To make the propeller boss, take the straight-grain pine and cut two lengths (8 inches [205mm] long, 2 inches [50mm] wide, and ¾ inch [20mm] thick); then use the pencil, ruler, and square to set out the lines to make cross lap joints and the angled propeller stubs. From one end to another, the measurements should be 2 inches (50mm), 1 inch (25mm), 2 inches (50mm), 1 inch (25mm), and 2 inches (50mm). Use the square to run the lines around the wood.

7 Use a flat chisel to cut the cross lap joints. Smooth the joints using the plane.

8 Push the two lapped pieces together to make the "X" shape of the propeller boss.

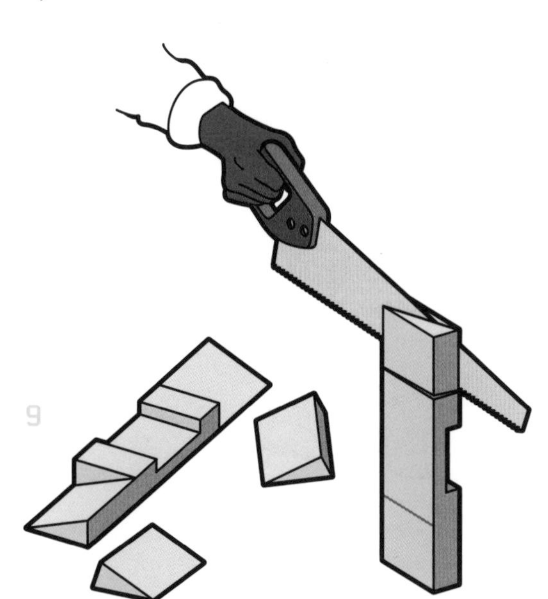

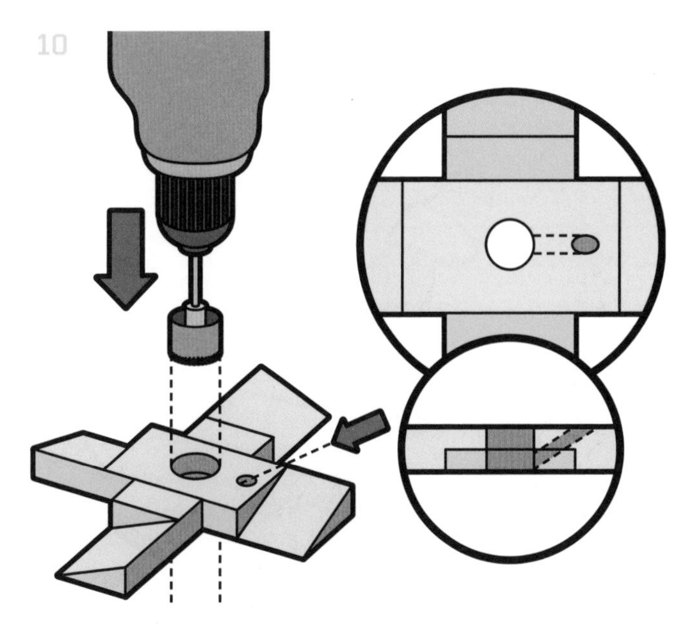

9 Use the straight saw to cut 45-degree angles into the left and right edges of both lapped pieces, creating four propeller stubs on each end. Make sure that they all angle in the same direction. Keep the little wedge-shaped pieces of waste.

10 Put the propeller boss together (it should be a tight fit) so that you have the "X" shape with all the stubs angled in the same direction, and use a drill fitted with a Forstner-type bit to bore a 1⅛-inch- (29-mm) diameter hole in the center of the boss. Drill a second hole at approximately 45 degrees through the boss (it should pass through both parts of the boss).

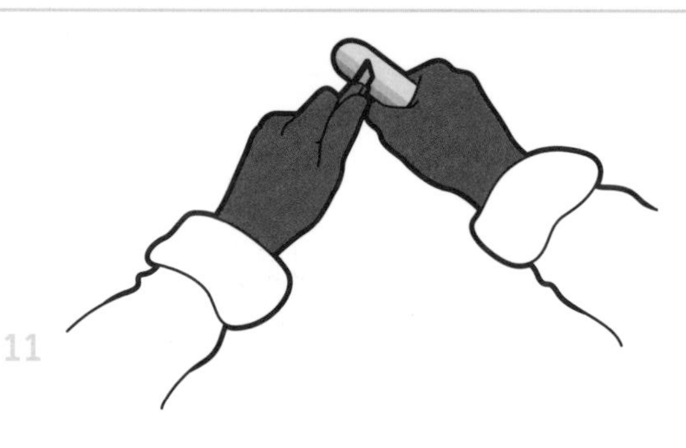

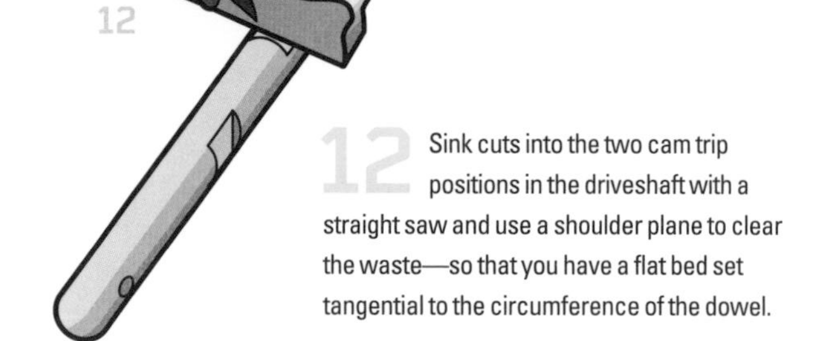

11 Cut the 1-inch (25mm)-diameter dowel into two pieces; a 12-inch (305-mm) length for the driveshaft, and an 18-inch (455-mm) length for the pivot pole. Take the driveshaft piece, mark out the position of the various holes and cam trips (see the template on page 158 for position and measurements), and use a craft knife and sandpaper to whittle and work the ends to a smooth 1-inch- (25-mm) radius finish.

12 Sink cuts into the two cam trip positions in the driveshaft with a straight saw and use a shoulder plane to clear the waste—so that you have a flat bed set tangential to the circumference of the dowel.

13 From cutoffs of pine, cut the two cam trips to size and length (3½ inches [90mm]) long, 1⅛ inches [30mm] wide, and ⅛ inch [3mm] thick). Use the plane and sandpaper to rub them down to a smooth round-edged finish. Nail the cam trips in place with brads on the prepared shaft.

14

14 Cut the two drumsticks to size, (5½ inches [140mm long], ⅝ inch [15mm wide]). Whittle and shape them in much the same way as described for the shaft, and then bind the drum-beating ends with 1-inch (25-mm) strips of lead or wire. On the opposite ends of each drumstick, drill a 5⁄16-inch- (8-mm) diameter hole around 1 inch (25mm) in from the end (for the wire to feed through).

15 Dry fit the pieces of the box together, gently hammering in as few brads as possible. Proceed slowly, constantly making sure that all edges and corners are well aligned.

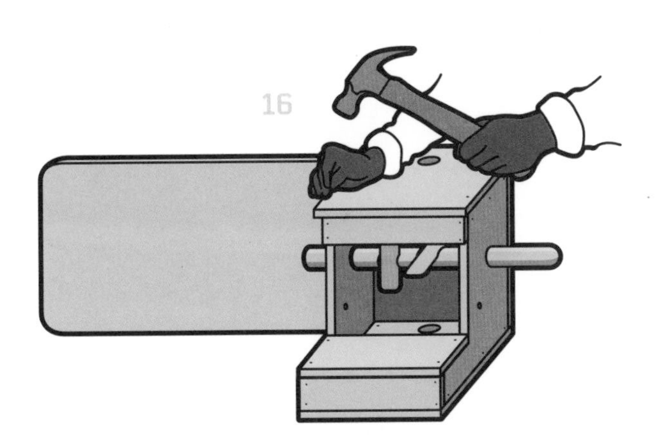

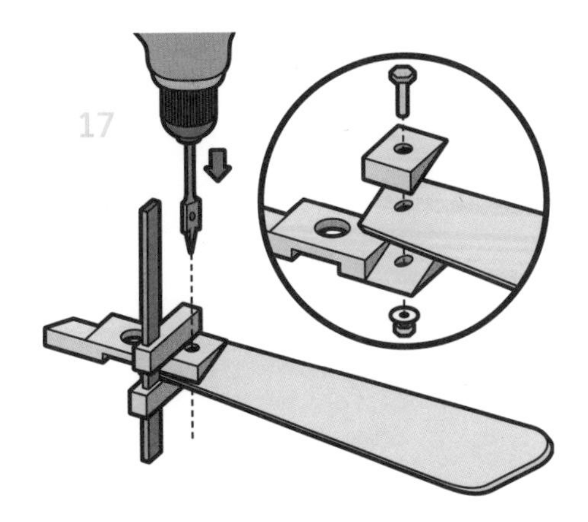

16 When you are happy with the fit, ease the box sides apart, slide the driveshaft in place, and hold the box together with bids.

17 One piece at a time, ensuring that each is angled to move counterclockwise, set a propeller blade on an angled end of a boss stub—the angle of each blade governs the direction of spin, so don't assemble them in the wrong direction. Top off each blade with a triangular wedge of waste, and secure the whole works with a clamp. Run it through with a ⅜-inch- (10-mm) diameter drill, and fix with a nut, bolt, and washer. Rerun this for all four blades.

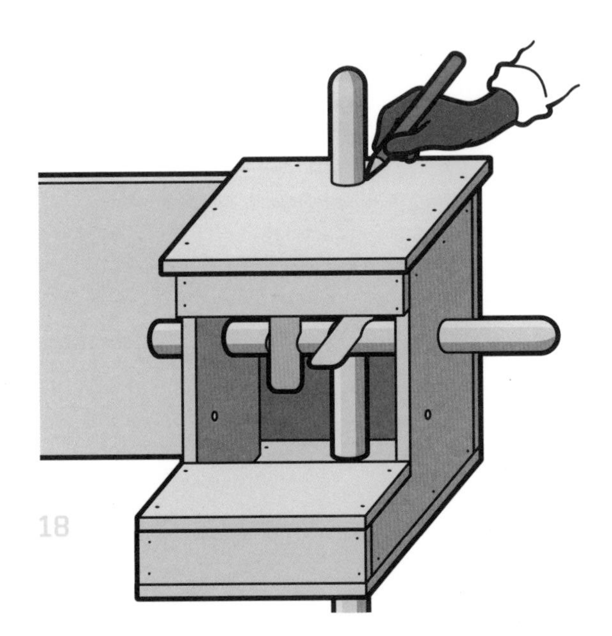

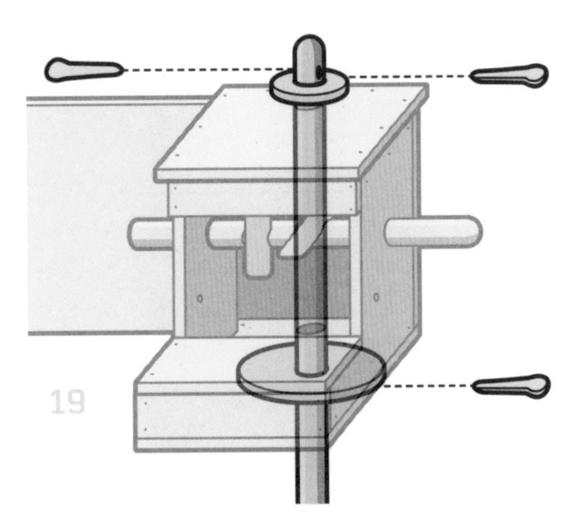

18 Take the 18-inch (455-mm) piece of dowel for the pivot pole and use sandpaper to rub each end down to a smooth, round-edged finish. Slide the box onto the pivot pole, allowing two inches of the pole to stand proud at the top. Mark in the position of the two peg holes on the pivot pole; one for the peg at the top of the pole, and a second for the peg supporting the base support disk. Remove the box and drill the necessary holes in the pivot pole.

19 Fit the large base-support disk over the pivot pole, secure this with one of the wooden pegs, and slide the box onto the pole, securing it at the top with a washer and two wedged pegs.

20 Fit two washers over the propeller end of the driveshaft and slide the propeller unit onto the shaft. Drill a hole into the drive shaft (using the existing hole in the boss as a guide) and use a wooden peg to hold the boss in place. Secure the other end of the driveshaft with a washer and peg.

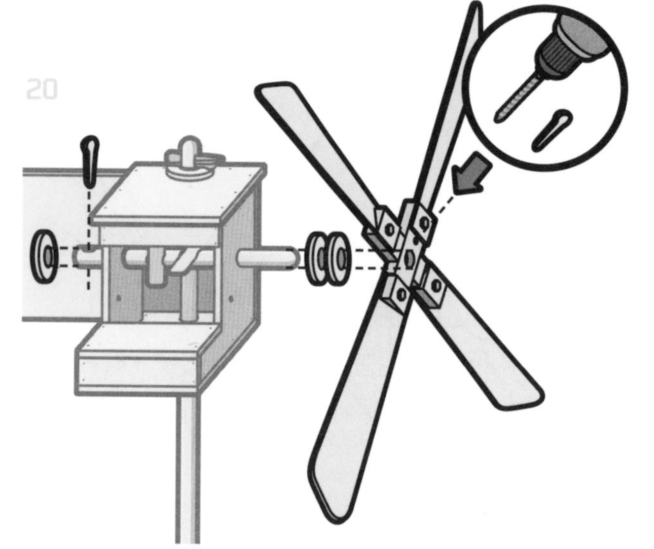

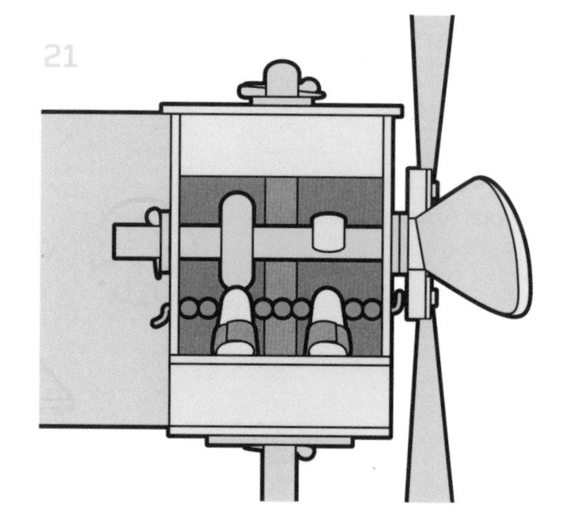

21 Feed the coat-hanger wire through one of the small holes on one side piece, slide and space the drumsticks along the length of the wire, arranging the spacers (we've used wooden beads) so that the rounded ends of the sticks are aligned and centered with the cam trips. Pull the wire through the small hole on the opposite side piece. Use pliers to bend and loop the ends of the wire outside the box, so that it stays firmly in place.

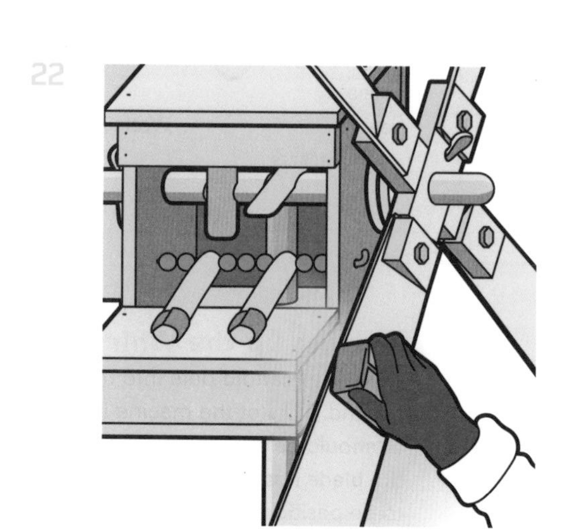

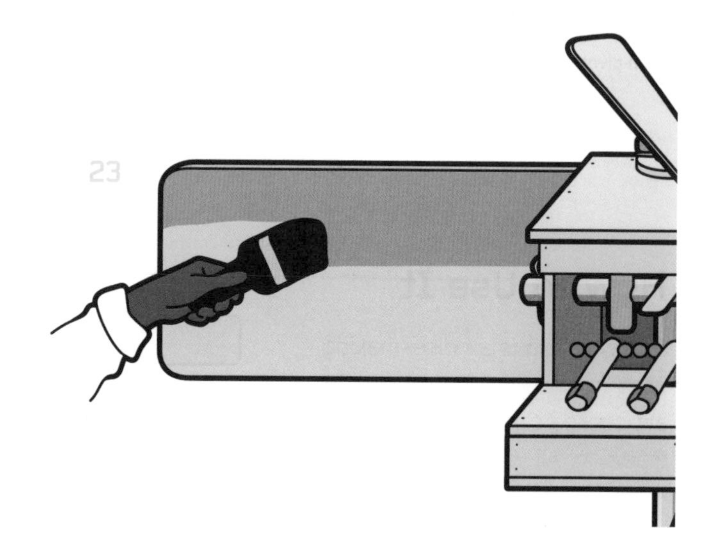

22 Adjust and sand the whole works so that the structure is smooth to the touch, and, if necessary, add more brads and screws to secure the unit.

23 Move to a dust-free area and lay on two coats of Danish oil and colors to your liking.

The Whirligig

The whirligig is exciting to watch in its inexorable function: the wind drives the sails, the driveshaft slowly turns, the cam trips press down on the drumsticks, and the noise is made.

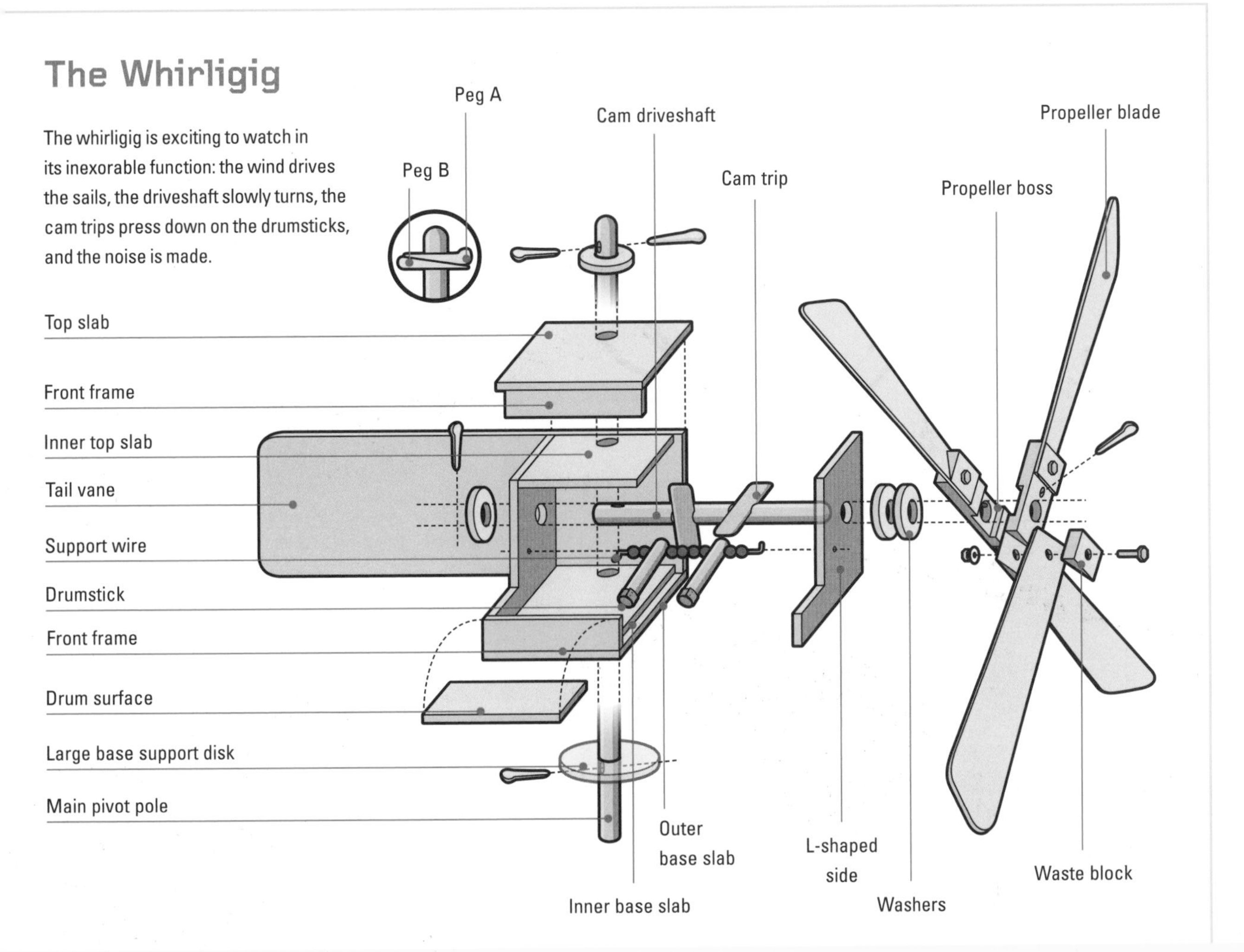

How to Use It

Our whirligig is a noise-making wind machine in the classic, folk art cuckoo clock and music box tradition. All you need to do now is position the scaffold pole, slot the whirligig into place, and while away the minutes looking and listening to the machine on a windy morning.

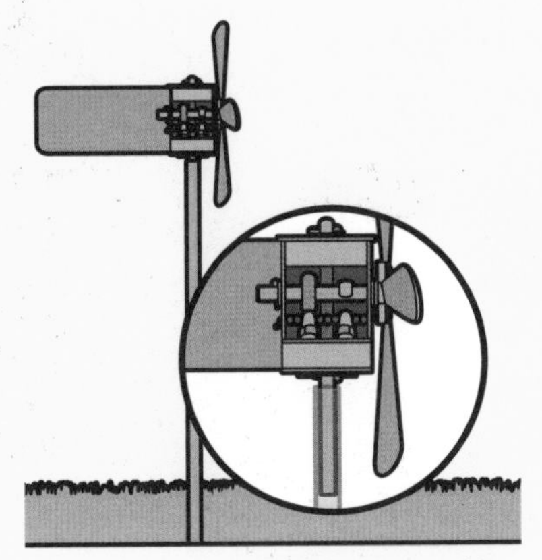

Slot the pivot pole into the top of the mast.

Setting up the Whirligig

Drive the scaffold pole into the ground and slot the macine into it (it should be a snug fit). Ensure that the blade tips are sufficiently long to be positioned above head height. Position the pole in an open space, and keep in mind the direction of prevailing winds, and the likelihood of people tripping over the pole.

Compact Theremin

Named after its Russian inventor, Léon Theremin, this electronic machine had a sinister start in life as a proximity detector device for the Soviet Secret Service. Despite its exotic roots, our version is purely a musical instrument. It operates using infrared light, using two transmitters; one to operate the pitch and the other the volume. It is "played" by passing your hands near to the transmitters to produce a note. Although this project calls for a substantial amount of circuit board work, it's straightforward; however, you should allow a weekend or two for its construction. With practice, we soon found that we could play full tunes in the theremin's singularly characteristic and somewhat doleful voice.

YOU WILL NEED

MDF to make up:

- A: piece, 8¾ × 10¼ × ¼ inches (225 × 260 x 6mm)
- B: piece, 6 × 10¼ × ¼ inches (150 × 260 x 6mm)
- C: piece, 2¼ × 10¼ × ¼ inches (55 × 260 x 6mm)
- D: piece, 5 × 10¼ × ¼ inches (125 × 260 x 6mm)
- E: piece, 5 × 10¼ × ¼ inches (125 × 260 x 6mm)
- F: piece, 8¾ × 6 × ½ inches (225 × 150 × 12mm)
- G: piece, 8¾ × 6 × ½ inches (225 × 150 × 12mm)
- H: piece, ½ × 10¼ × ¼ inches (12 × 260 × 6mm)
- Four ¾-inch (20-mm) screws, ⅛-inch (4mm) in diameter, and nuts
- Twelve ½-inch (12-mm) no. 4 wood screws
- 10-foot (3-m) length of thin PVC insulated wire (size not critical)
- 40 inches (1,000mm) of 22-AWG-(0.75mm diameter) tinned copper wire (for assembling components on the circuit board)

Electronic components:

- 0.1-inch (2.54-mm) pitch matrix circuit board, 5 × 5 inches (125 × 125mm)
- AA batteries (or any of the simple 12 volt units that plug into a 13A socket)
- 220R ½ W resistor (R1)
- 220R ¼ W resistor (R2)
- 470R ½ W resistor (R5)
- 4K7 ¼ W resistor (R3, R7)
- 2K2 ¼ W resistor (R4)
- 470K ¼ W resistor (R6)
- 10R 1 W resistor (R8)
- 1M ¼ W resistor (R9)
- 4.7 nF capacitor (C1)
- 100 uF 25 v electrolytic (C2)
- 2N 7000 MOSFET (Q1, Q2)
- BD681 Darlington transistor (Q3)
- IR emitter diode, Sharp PD410PI (D1, D3)
- IR photodiode, Osram SFH4505 (D2, D4)
- LM358N, Dual operational amplifier
- 4011, Quad NAND gate
- 8-ohm loudspeaker, nominal 4 × 7 inches (100 × 180mm), elliptical

TOOLS

- Electric drill/driver with a good selection of drill bits to fit
- Hacksaw or saber saw
- Hole cutter, ¼ × 1 inch (5 × 25mm) (optional, for loudspeaker aperture)
- Pliers
- Sandpaper and sanding block
- Screwdriver
- Side cutters
- Soldering iron (with cored solder and flux)
- White glue

How to Make the Compact Theremin

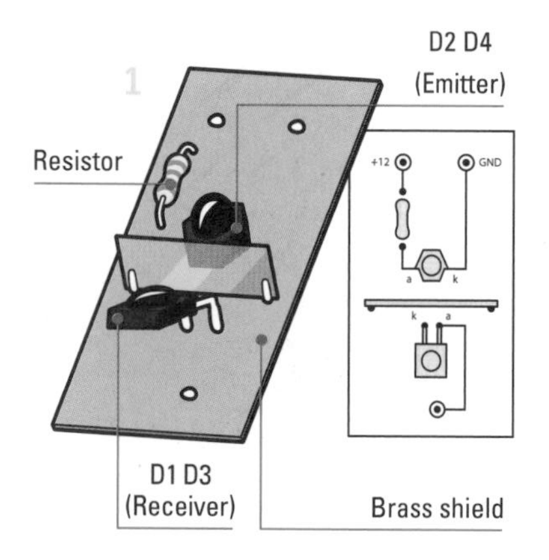

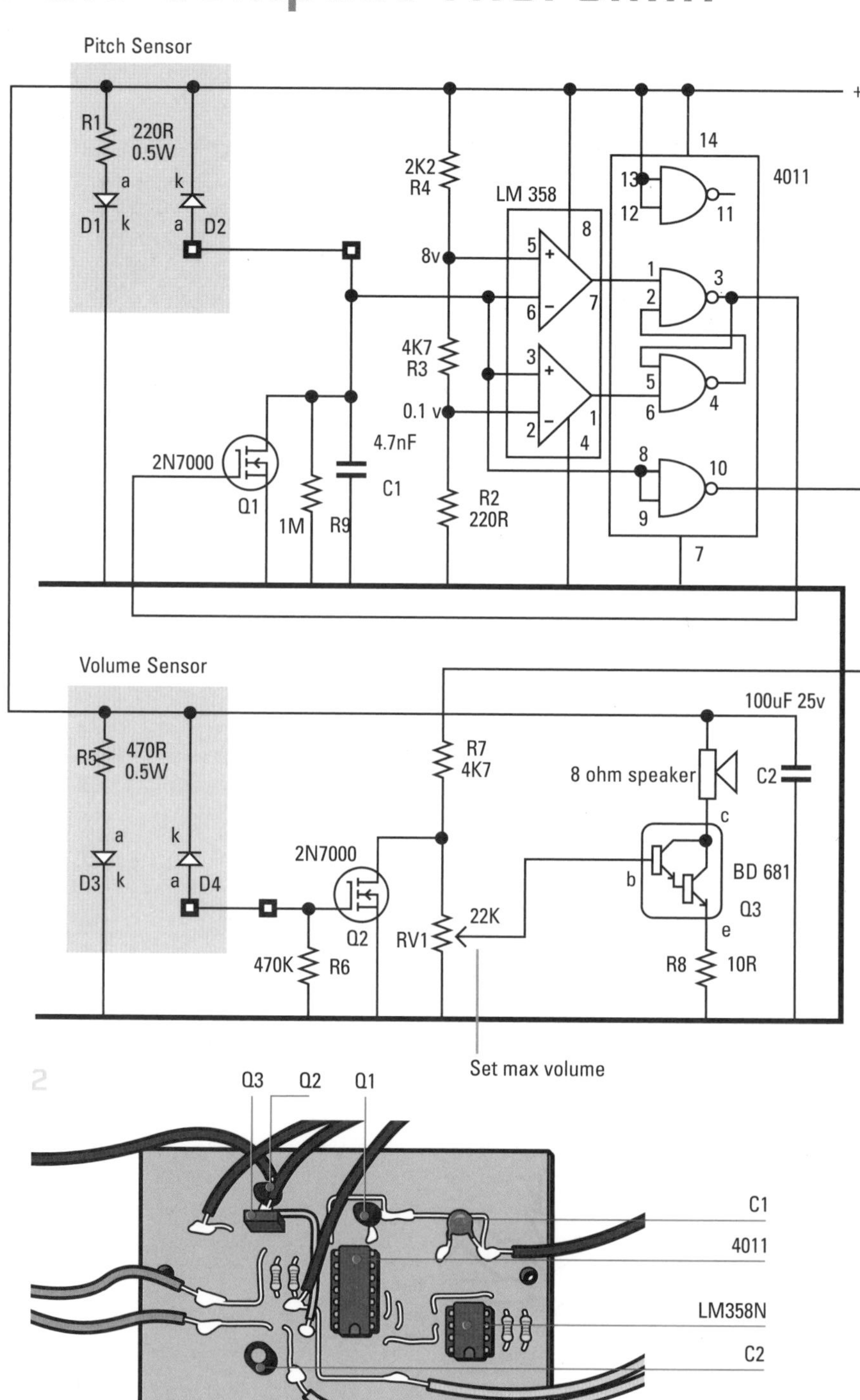

1 Cut two 1 x 2-inch (25 x 55-mm) pieces from the 0.1-inch (2.54mm) pitch matrix board using a hacksaw. Using the pliers and side cutters to trim the components and lengths of copper wire, construct the electronics on two separate and identical IR sensor boards, following the circuit diagram shown. The exact layout is not critical, but make certain that the diodes are connected correctly, because it is important to shield the direct path from the emitter diode to the receiver diode. Also note that the receiver is mounted on two brads and angled at about 45 degrees.

2 Next, cut one 4 x 2½-inch (100 x 65-mm) piece of pitch matrix board, and assemble the components on the main circuit board, following the circuit diagram shown. The setting of the volume sensor is not so critical, with the exception of the emitter current, which is half the current used in the pitch sensor.

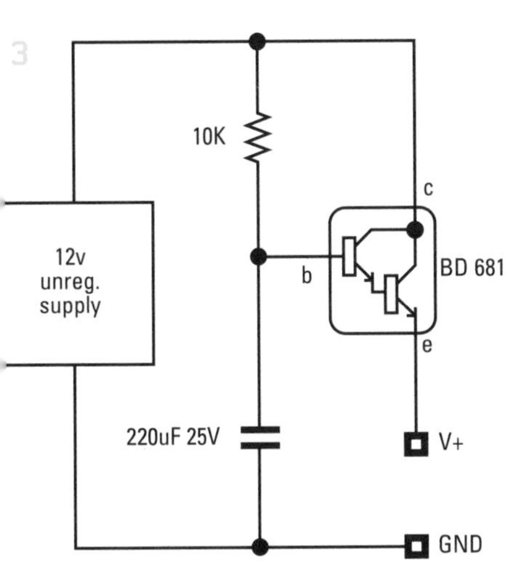

3 The theremin is designed to run from a nominal 12v supply, although 8v is also acceptable. Use AA batteries or any of the simple units that plug into a 13A outlet for a smooth operation, with no hum or growling sound (caused by ripple on the supply). If you are using a non-regulated supply, the audio quality is likely to be degraded. To avoid this, add a simple regulator (a hum-removing module), as shown in the circuit diagram above.

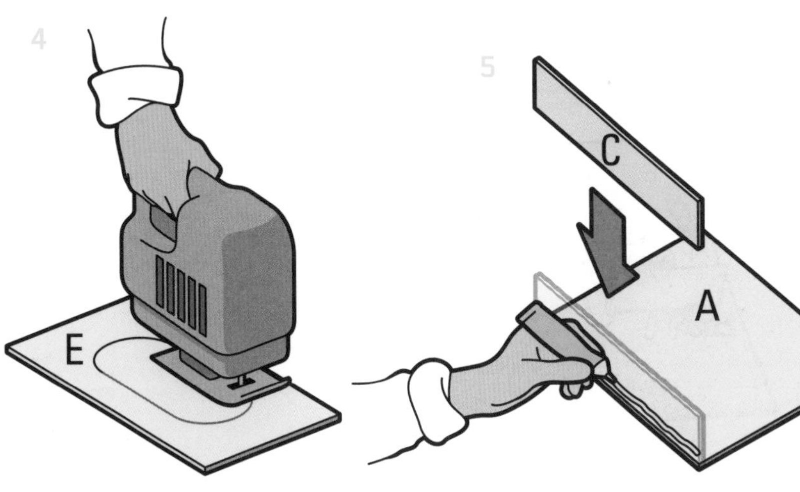

4 To make the cabinet, use a saber saw to cut the following pieces from the ¼-inch (6-mm) MDF: piece A (for the bottom piece), 8¾ × 10¼ inches (225 × 260mm); piece B (for the back piece), 6 × 10¼ inches (150 × 260mm); piece C (for the front piece), 2¼ × 10¼ inches (55 × 260mm); piece D (for the top piece), 5 × 10¼ inches (125 × 260mm); and piece E (for the loudspeaker piece), 5 × 10¼ inches (130 × 260mm). Cut a hole to match your loudspeaker in piece E.

5 Glue front piece (C) to the edge of the bottom piece, as shown. Make certain it is at a right angle to the bottom piece.

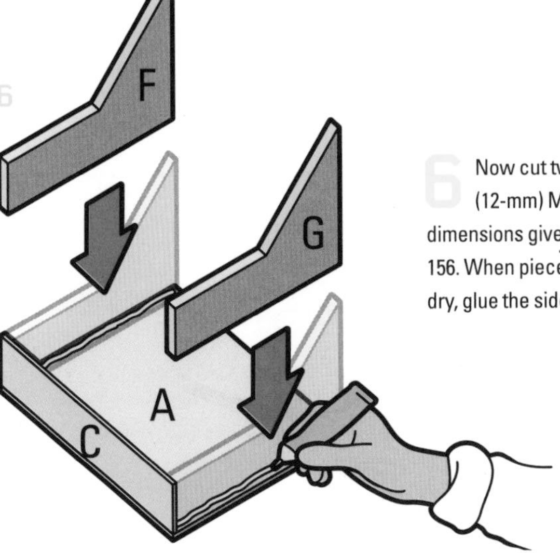

6 Now cut two pieces of the ½-inch (12-mm) MDF, to the shape and dimensions given in the template on page 156. When piece C and the bottom piece are dry, glue the side pieces F and G in position.

7 Attach your loudspeaker to the loudspeaker piece (E) using four ¾-inch- (20-mm) long screws and nuts, ⅛-inch- (4mm) in diameter, and rest piece E in the final location on the side pieces. The speaker magnet will possibly project past where the back piece will be located, so you may need to cut a hole in back piece B for clearance.

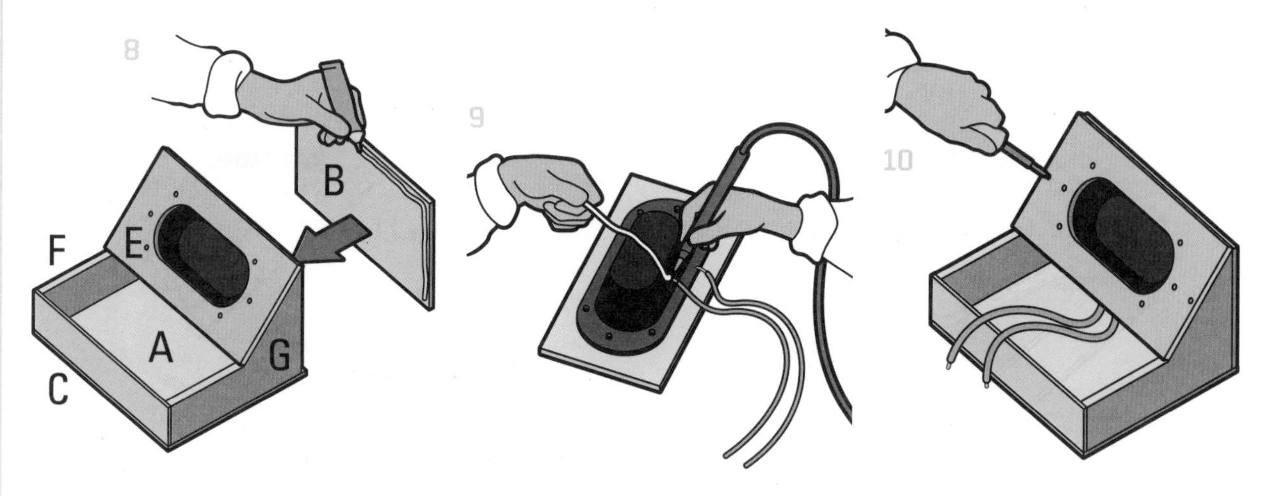

8 When trimmed to the correct size, the back piece (B) can be glued in place.

9 Solder wires approximately 12 inches (300mm) long to the loudspeaker.

10 Fix the loudspeaker piece (E) in place using six ½-inch (12-mm) no. 4 wood screws.

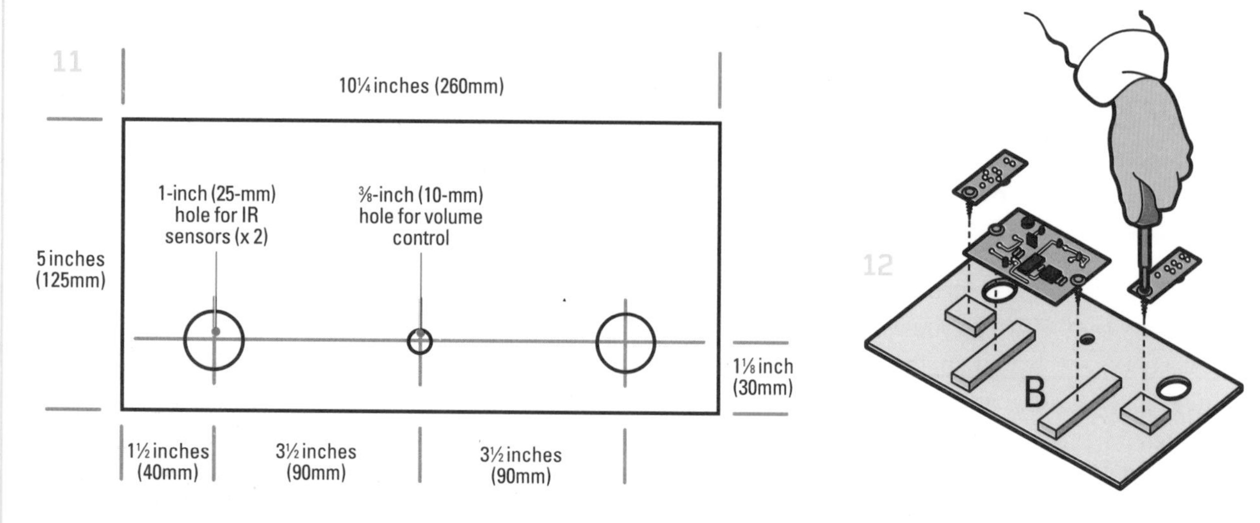

Use the saber saw to cut holes in top piece (D), as shown in the illustration.

12 Use the cut off wood pieces from the ½-inch (12-mm) MDF to cut two strips and two blocks, as shown. The dimensions of these are not critical. Mount the main circuit board onto the two strips, and the two IR sensors on to the two blocks. Attach these to the underside of the top piece using four ½-inch (12mm) no. 4 wood screws (two to secure the circuit board, and one to attach each of the sensor boards). The location of the main circuit board is not critical, but the two IR sensors must be held so that the two diodes are centered in the 1-inch (25-mm) holes.

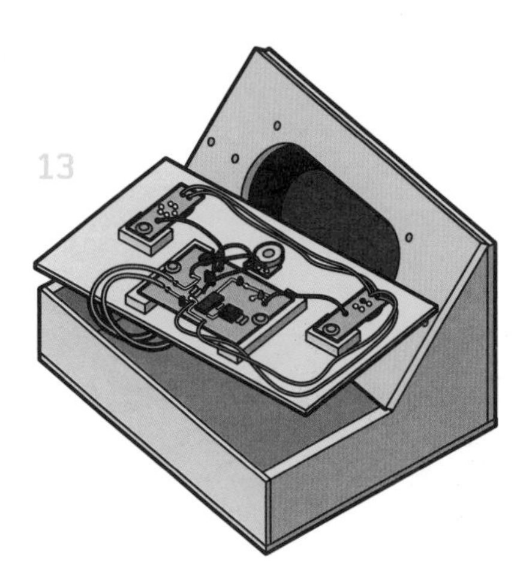

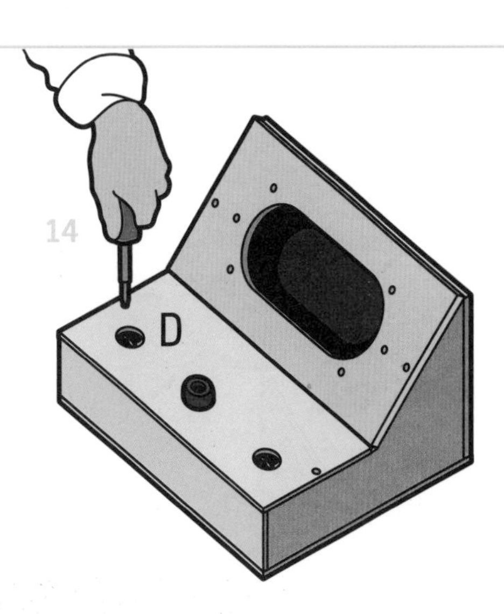

13 Wire the loudspeaker and two IR sensors to the circuit board and connect the power supply. If the hum removing module is added (see step 3), secure it on another ½-inch (12-mm) block of MDF glued to the bottom piece, clear of any of the other components.

14 Attach the top piece (D) with two ½-inch (12mm) screws into the side pieces F and G. Glue strip H to the top of the theremin where pieces B and E meet.

How to Use It

The machine is now ready for its first tune. Simply sweep your thumbs gently from side to side over the sensor area, and you will hear its melancholy voice. The left thumb controls the volume; the right alters the pitch. Perhaps you would like to create an electronic orchestra? If so, the didgeridoo on pages 116–123 will make a suitable partner if you haven't made it already. Their voices are perhaps competitive rather than complementary, but you can have a lot of fun seeing which will outplay the other.

Altering the Pitch and Volume

The angle at which the photodiodes in the infrared sensors is set will alter the sound of the instrument. You can fine-tune the angle to give the maximum frequency change as the hand or thumb approaches by setting it to maximum reflection. The sensitivity of the volume control is determined by R7, a 470K ohm resistor—reduce the value of the resistor if you feel the change of volume is too abrupt.

Tip

Do not play the machine in bright sunlight, because although the theremin operates with infrared light, natural strong sunlight can produce sufficient infrared light to produce an output signal.

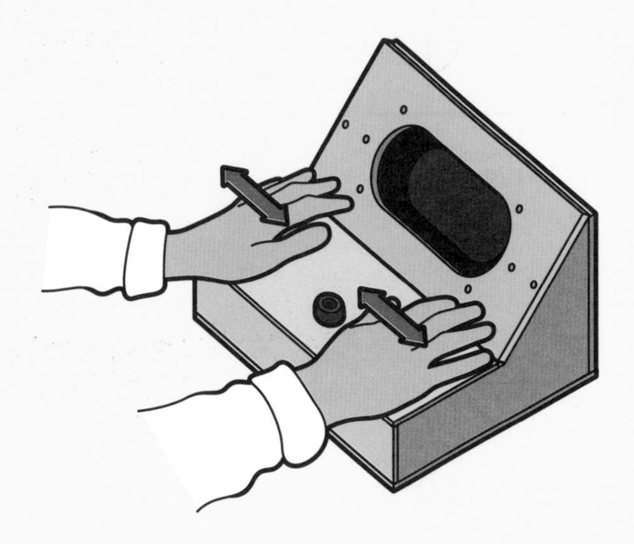

C
D
E
F
G
A
B

Electronic Music Box

Welcome to our electronic version of a traditional music box. Instead of steel pins, it uses old CDs with holes drilled through them. A light shines above the CD and a corresponding set of photodiodes is placed underneath it; as the CD turns, the light shines through the holes onto the photodiodes, creating currents that, in turn, choose resistors,which provoke oscillators to produce the desired notes.Even if the electronic explanation sounds too technical for you, don't be put off. There's something appealing about a machine that produces a traditional result by entirely contemporary means. The number of tunes you can create is limited only by your own ingenuity and a steady supply of redundant CDs.

YOU WILL NEED

Plywood to make up:

- A: piece, 4¼ × 5 × 5/16 inches (110 × 130 x 8mm)
- B: piece, 4¼ × 5 × 5/16 inches (110 × 130 × 8mm)
- C: piece, 3½ × 5 × 5/16 inches (90 × 130 × 8mm)
- D: piece, 3½ × 5 × 5/16 inches (90 × 130 × 8mm)
- E: piece, 4¼ × 4⅝ × 5/16 inches (106 × 110 × 8mm)
- F: piece, 4¼ × 4⅝× 5/16 inches (106 × 110 × 8mm)
- G: piece, 3½ × ⅝ x 5/16 inches (90 × 15 x 8mm)
- H: piece, 3½ × ⅜ x 5/16 inches (90 × 10 x 8mm)
- I: piece, 7 × 8⅝× 5/16 inches (180 × 220 × 8mm)

Medium-density fiberboard (MDF) to make up:

- J: piece, 3 × 1⅛ × ¼ inches (75 × 30 × 6mm)
- L: piece, 1⅛ × 1⅛ × ¼ inches (30 × 30 × 6mm)
- M: piece, 2⅜ × ¾ × ¼ inches (60 × 20 × 6mm)
- N: piece, 2 × 2 × ½-inches (50 × 50 × 12mm)—allows for waste
- O: two pieces, 1⅛ × 2 × ½-inches (30 × 50 × 12mm)
- P: piece, 2 × 1 × ½-inches (50 × 25 × 12mm)
- Q: piece, 2⅜ × ⅝ inch × 5/16 inches (60 × 15 × 8mm)
- R: piece, 1⅛ × ⅝ × ⅛ inches (30 × 15 × 3mm)
- Two pieces of sponge rubber, 2 × ½ × ⅛-inches (50 × 12 × 4mm)
- Two ¾-inch (19-mm) no. 4 wood screws
- 2 inch (25mm) of ⅝-inch-diameter (15-mm) dowel—allows for waste
- 6½ feet (1,980mm) of 7/0.25-mm-diameter PVC-insulated hook up wire
- 40 inches (990mm) of 22-AWG-(0.644-mm diameter) tinned copper wire
- Eight ½-inch (12-mm) no. 4 wood screws
- Piece black plastic, 1⅛ x ⅝ x 1/32 inches (30 x 15 x 1mm)
- Two PP3 battery clips

Electronic components:

- 0.1-inch (2.5mm) pitch matrix board, 4 × 6¼-inch (100 × 160mm)
- CD drive motor (a range of units is available, all running from 1.5 to 3v; as long as the output speed is around 1RPM or slightly faster, any unit may be used)
- 220K ¼ W (R1)
- 100K ¼ W (R2)
- 470K ¼ W (R3)
- 2K2 ¼ W (R4)
- 1M0 ¼ W (R5–R12)
- 10R ¼ W, located on illuminator assembly
- 470K preset, frequency adjusting variables (RV1–RV)
- MC33201, operational amplifier (IC1)
- 74HC03, quad open drain (IC2, IC3)
- BD681, Darlington transistor (Q1)
- SFH 229, photodiode (D1–D8)
- 1N4001, diode (D9, D10)
- 47nF (C1, C2)
- 100nF (C3)
- YH70 M, Infrared emitter (Da Db)
- Two battery holders (each for two AA batteries)
- Four AA batteries
- SPDT switch (SW1, SW2)
- 8-ohm loudspeaker, 3-inch diameter (8R)

How to Make the Electronic Music Box

TOOLS

- Electric drill, with a good selection of drill bits to fit
- Hacksaw or saber saw
- Ink pen with a fine point
- Impact adhesive
- Musical instrument (such as a piano or recorder), tuning device, or frequency meter, to set the notes
- Protractor
- Sandpaper and sanding block
- Side cutters
- Small long-nose pliers
- Soldering iron (with cored solder and flux)
- Two-tube pack of epoxy; one tube is the resin, the other is the hardener
- White glue

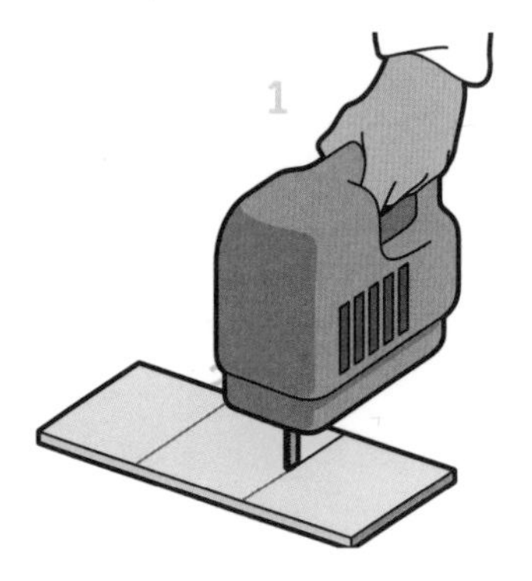

1 To build the cabinet that houses the main parts, including the motor gearbox, circuit board, photodiode assembly, and loudspeaker, use the saber saw to cut the following pieces of 5/16-inch (8-mm) MDF: pieces A and B (side pieces), 4¼ × 5 inches (110 × 130mm); pieces C and D (top and bottom pieces), 3½ × 5 inches (90 × 130mm); pieces E and F (front and back pieces), 4¼ x 4⅝ inches (106 × 110mm); piece G (front support) 3½ × ⅝ inches (90 × 15mm); and piece H (inner piece), 3½ × ⅜-inch (90 × 10mm).

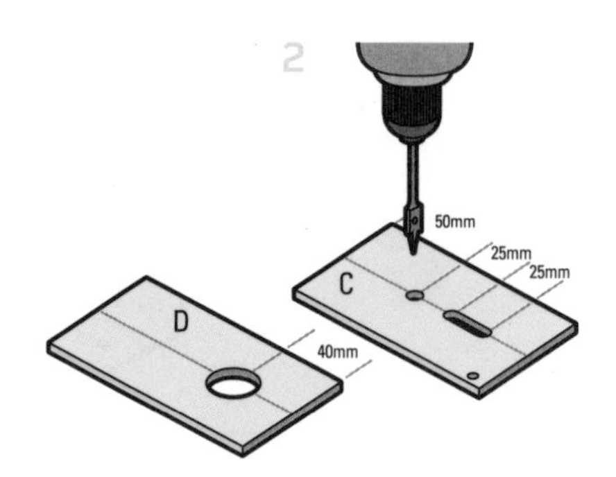

2 In the top of piece C, drill a centered ⅜-inch (11-mm) hole for the gearbox spindle, 2 inches (50mm) from one edge; then, leaving a gap of 1 inch (25mm), cut a centered slot for the photodiodes, 2⅛ inches (55mm) long and ⅜-inch (11-mm) wide. In the near corner, drill a ¼-inch (6-mm) hole for the illuminator wires. In part D, drill a 1⅛-inch (30-mm) hole in the centerline, 1½ inches (40mm) from one of the short sides.

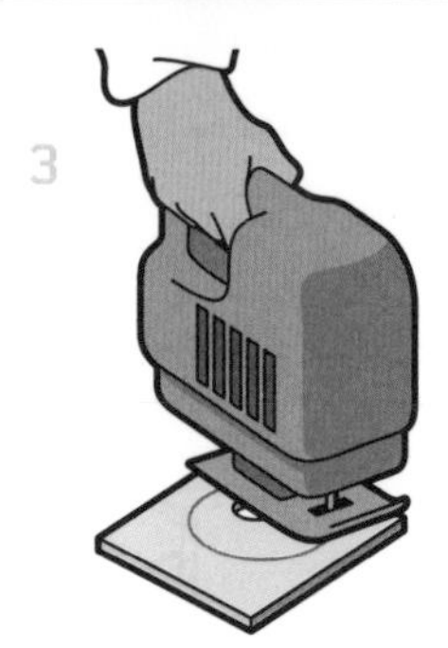

3 Use the saber saw to cut a circular hole in piece F to accommodate the loudspeaker.

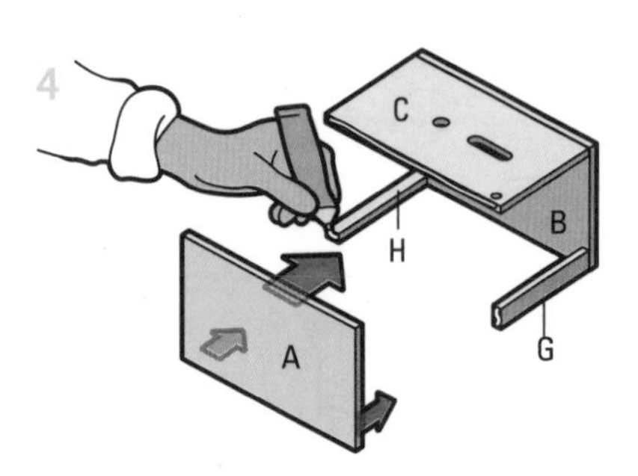

4 Glue pieces A, B, C, G, and H together, as shown above right, to make a box.

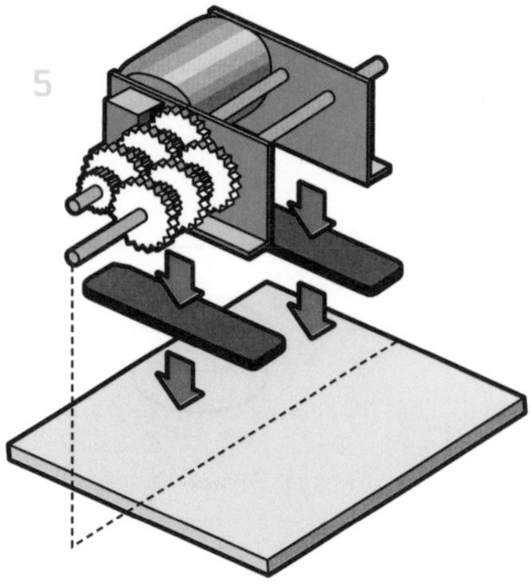

5 Next fit the motor/gearbox to the front panel E, positioning it so that the output shaft lies on the centerline. Do not use the slots provided in the motor unit for the screws. If you do so, when the motor rotates, it will use the panels as a sounding board and the mechanical noise will be greatly magnified. To avoid this, mount the motor, as shown, on two 1/6-inch- (4-mm) thick sponge rubber pads, glued with impact adhesive. Ensure that piece E slides smoothly into the box.

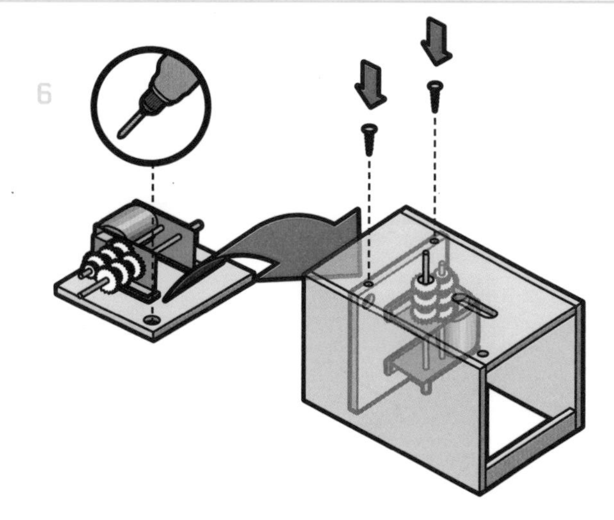

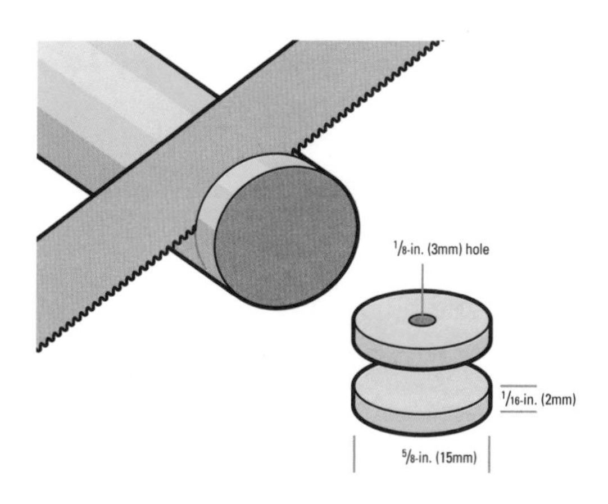

6 Drill a 3/8-inch (11-mm) hole (for the circuit board) in the top corner of piece E. Apply white glue to the top and side edges and then attach piece E inside the box, positioned so that the output shaft of the motor/gearbox is centered in the 3/8-inch (11-mm) hole on the top piece.

7 Next comes the drive wheel, which couples the CD to the output shaft of the motor/gearbox. Use the hacksaw to cut a small disk of dowel, 5/8-inch (15mm) in diameter (the size of the center hole in a CD), and 1/8-inch (3mm) thick. Drill an 1/8-inch (3-mm) hole through this, concentric with the circumference. (If your dowel is larger than 5/8 inch (15mm) in diameter, you can use the "poor man's lathe" by drilling an 1/8-inch (3-mm) hole through it, putting a 1/8-inch (3-mm) screw through the hole, and placing this in the chuck of an electric hand drill. Carefully sand it to the correct diameter.) Make a second circle of dowel, the same size as above (this is needed for building the jig).

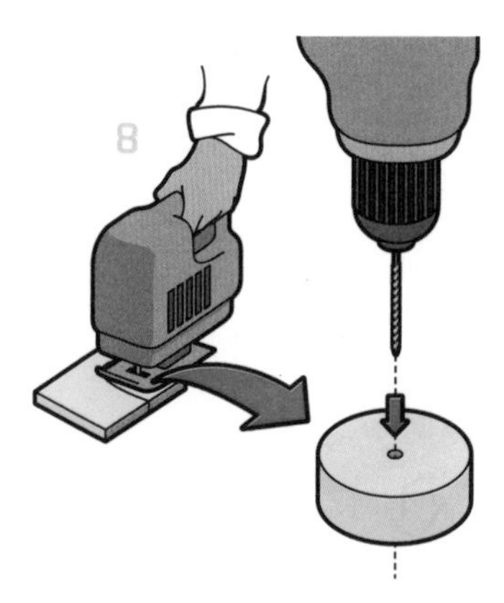

8 Now make a larger 1 3/8-inch- (35-mm) diameter circle by cutting piece N to an approximate disk shape (using a saber saw). Drill a 1/8-inch (3-mm) hole in the center, sanding to a smooth finish (the diameter of this part is not critical; it just serves as a resting place for the CD).

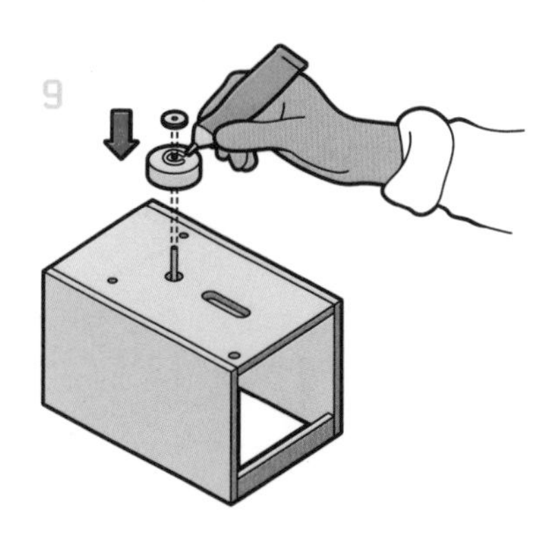

9 Glue the two disks together with impact adhesive, and push them into place on the output shaft.

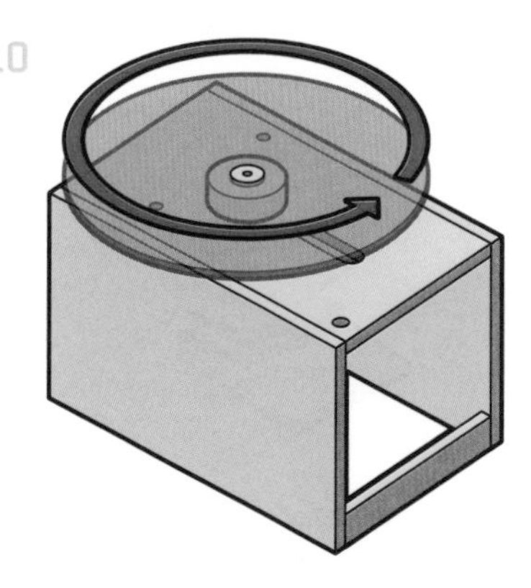

10 Power up the motor and check that the CD rotates smoothly (it should not wobble up or down or side to side, because this will affect the correct path of light through the drilled holes onto the photodiodes (fitted later).

11

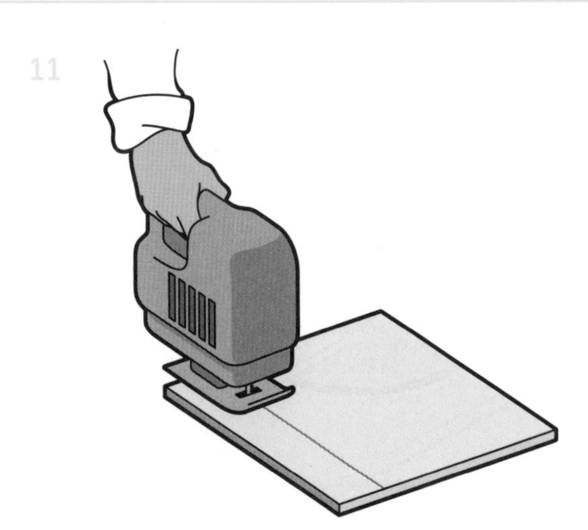

12

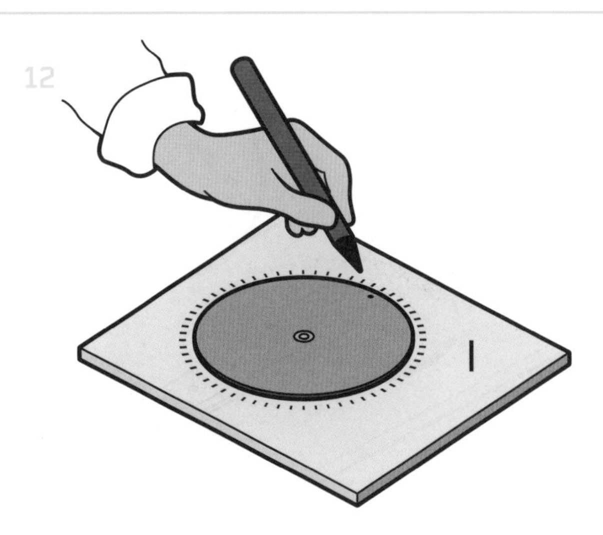

11 A simple jig is needed to drill correctly positioned holes later in the CDs. To make the jig, use a saber saw to cut the following: piece I (jig base), 7 × 8⅝ inches (180 × 220mm) from plywood; piece J (jig arm), 3 × 1⅛ inches (75 × 30mm) from MDF; piece K (locating spindle), is the second ⅛-inch- (2-mm) thick, ⅝-inch- (15-mm) diameter piece of dowel cut in step 7; and piece L (spacer), a 1⅛ x 1⅛-inch (30 x 30-mm) piece of ¼-inch (6-mm) MDF.

12 Screw piece K (the locating spindle) to the center of piece I (the base) with a ½-inch (12-mm) no. 4 wood screw. Make a black, ink reference mark on a CD, then place the CD on to K. Mark a point on the base just outside the edge of the CD and then use a protractor to add another 59 marks around the CD at precise 6-degree intervals (you may want to drill 1/32-inch (1-mm) holes in these afterward). These will enable the "programmer" or "composer" to rotate the CD in predetermined steps when creating the final tune.

13

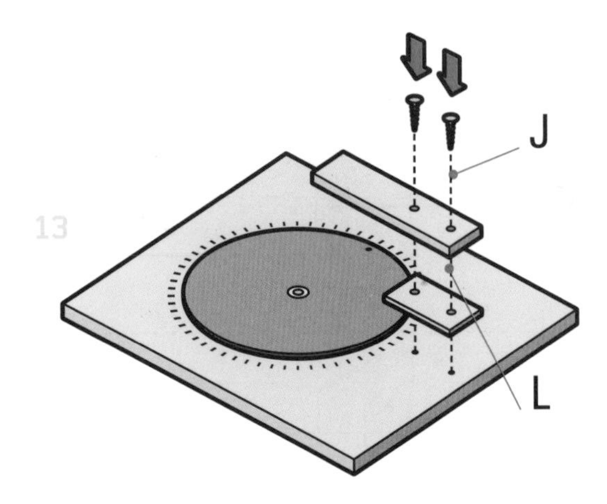

14

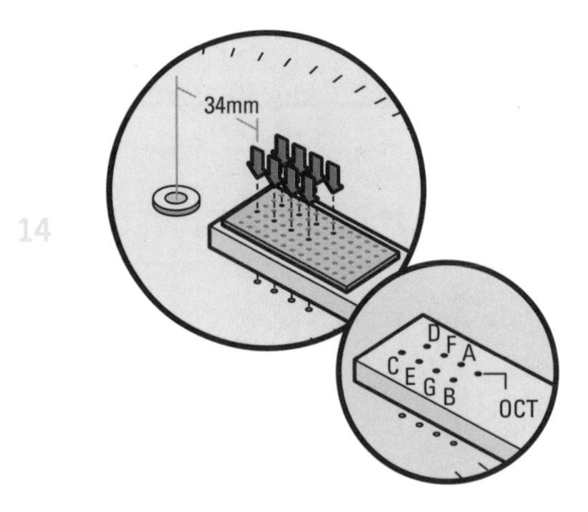

13 Place the spacer centered along the right edge of the jig base, position the jig arm over this, drill two holes for the screws, and use two ¾-inch (19-mm) no. 4 wood screws to screw both to the base.

14 Cut a 1½ x ¾-inch (37 x 18-mm) piece of matrix board with a hacksaw. Position it on the inner end of the jig arm and drill through the holes in the matrix (alternately) as shown, ensuring you drill through both the jig arm and base. Use a pen to label the notes, as shown.

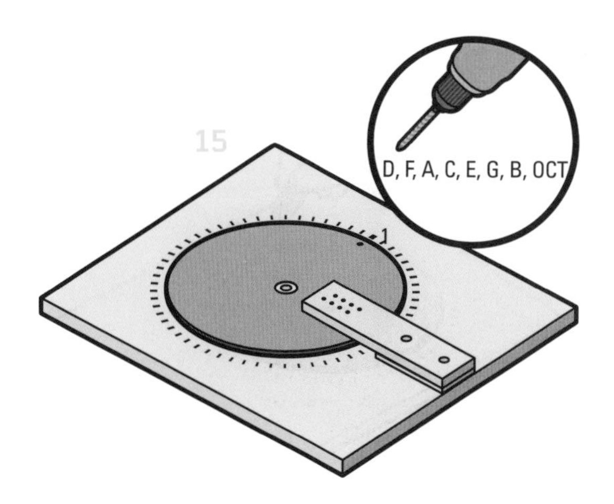

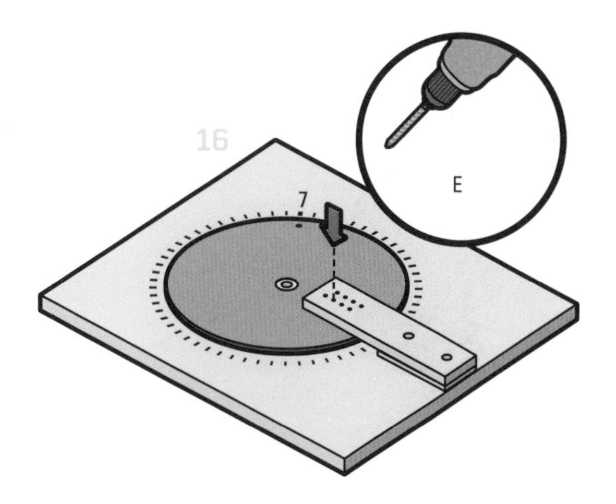

15 To enable the machine to be assembled and tested, you now need two sets of holes in a test CD. The first group will be at "time slot 1." Insert the CD, label side up, into the jig and set the black marker dot to 12 o'clock. Now, using a 1/16-inch (1.5-mm) drill bit (do not attempt to drill through the CD at this stage), lightly mark the CD through all eight hole positions.

16 Rotate the CD to position 3 (counter-clockwise) and lightly mark the disc through the C hole. Continuing counter-clockwise each time, continue rotating and lightly marking the CD as follows: position 5, through the D hole; position 7, E hole; position 9, F hole; position 11, G hole; position 13, A hole; position 15, B hole; position 17, octave hole. Finally, keeping the CD at position 17, lightly mark through the C hole.

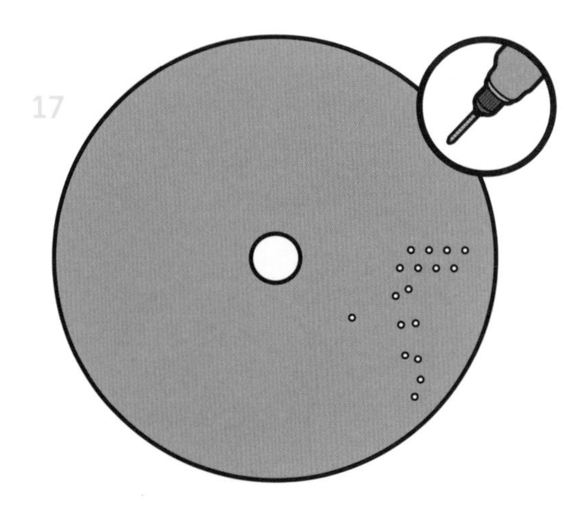

17 Remove the CD from the jig and with a 1/12-inch (2-mm) drill bit, using the marked holes, drill carefully right through the CD. Remove the burrs on the second set of holes with a 1/4-inch (6-mm) drill bit. The data on the CD can be presented in tabular form as shown in the table on the right.

Time slot	C	D	E	F	G	A	B	Shift octave
1	•	•	•	•	•	•	•	
2								
3	•							
5		•						
7			•					
9				•				
11					•			
13						•		
15							•	
17	•							•

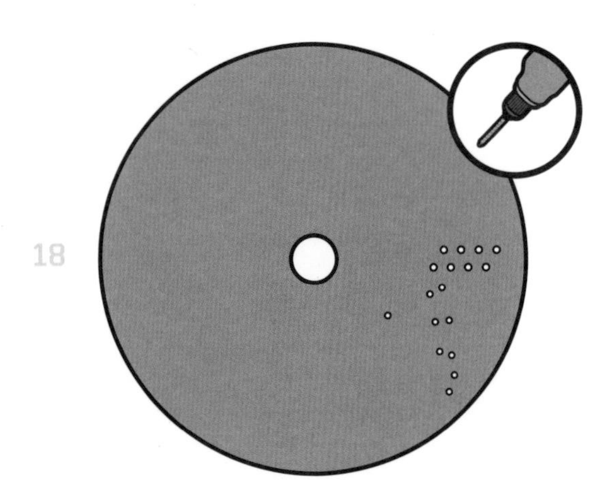

18 Using a 1/8-inch (3-mm) drill bit, open up the first group of holes (two rows of four aligned) to 1/8-inch (3-mm) holes. These will be used to line up the eight photodiodes in the finished machine. Again, remove the burrs in the holes. The critical operations are now over.

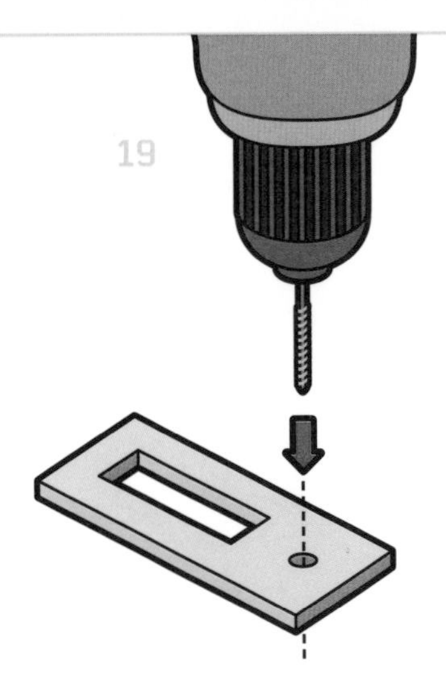

19 To prepare the photodiode array, first cut part M, 2 3/8 inches × 3/4 inches (60mm × 20mm), from 1/4-inch (6-mm) MDF. Include a 3/8-inch- (11-mm) wide cutout for the wires to exit the box through the slot on the upper plate. Drill a 1/4-inch (6-mm) hole, as shown, to permit the finished assembly to be screwed to the upper surface when positioned correctly below the CD with the group of eight test holes.

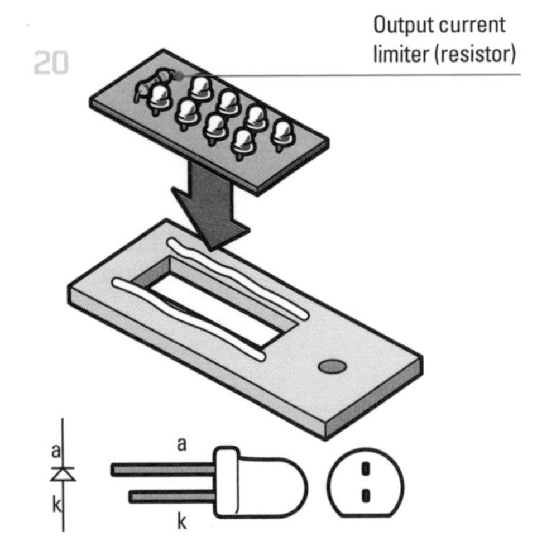

20 Cut a 1 3/8 × 5/8-inch (35 x 15-mm) piece from the 1/16-inch (2-mm) pitch matrix board using a hacksaw. Mount the eight diodes and the 470R resistor (which limits the output current) on the board. Epoxy the mount to the MDF base. (The diodes are very small, so be careful to ensure they are wired the right way around. Each diode has one wire longer than the other. A flat is also located on the diode package on the cathode side, but because the package is clear plastic, you will need to look closely to see the difference.)

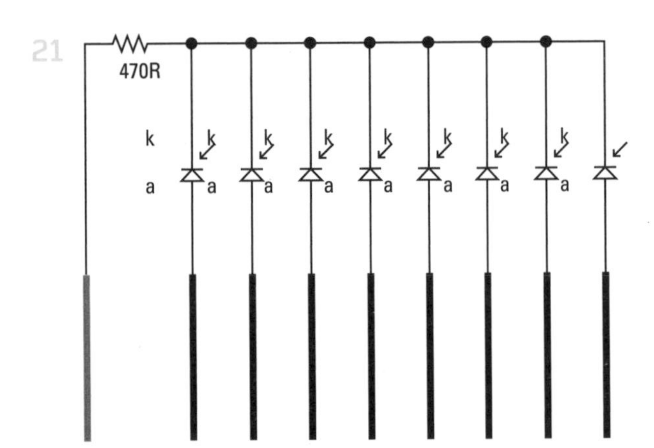

21 Cut the PVC-insulated cable into eleven 7-inch- (180-mm) long lengths, reserving two for securing the speaker. Attach the remaining seven lengths to the photodiode array, whose circuit is shown above. These cables will feed through the 3/8-inch (11-mm) hole next to the motor/gearbox and terminate in the circuit board compartment.

22 To make the main circuit board, cut a 3½ x 4-inch (87.5 x 100-mm) piece from the 1/16-inch (2-mm) pitch matrix board using a hacksaw. Using the pliers and side cutters to trim the components and lengths of copper wire, construct the electronics, following the circuit diagram shown on the left and the diagram on the right. When connecting the photodiodes to the circuit board, place them in musical order, as shown.

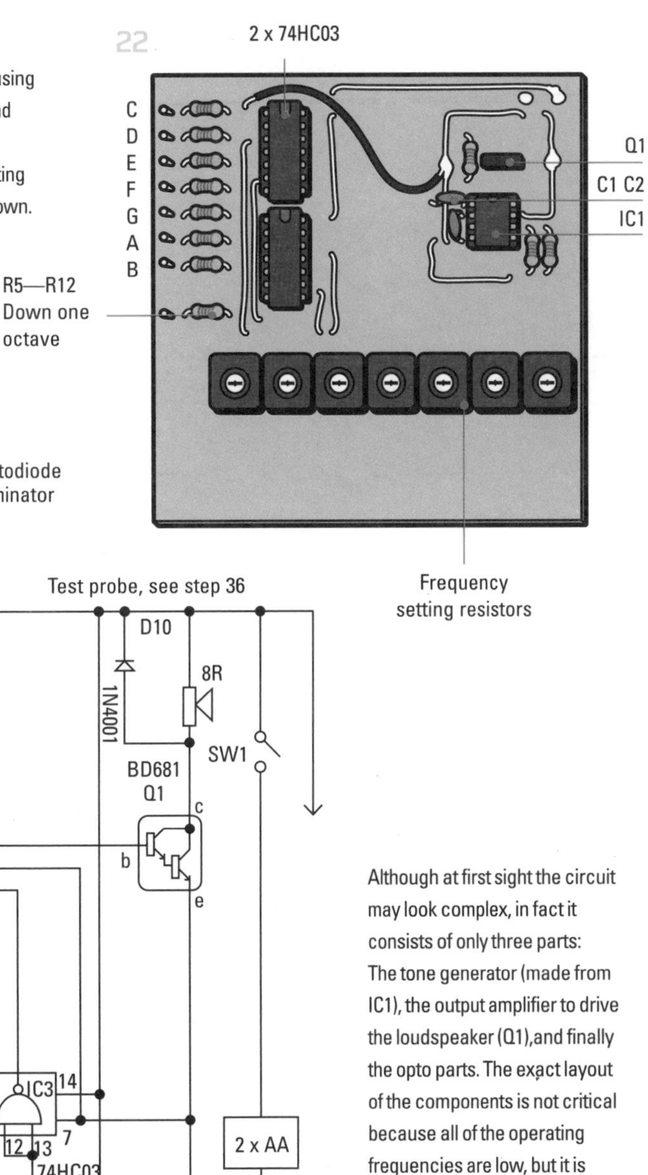

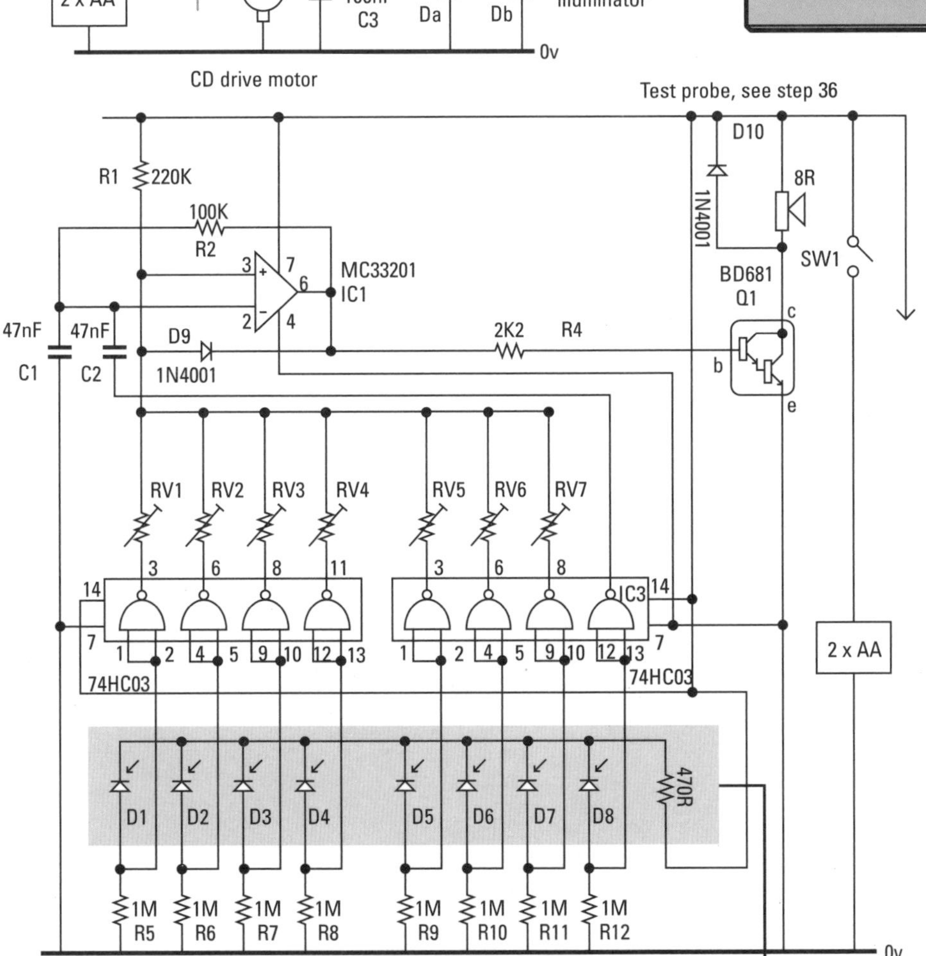

These components are located on the upper surface of the machine under the CD.

Although at first sight the circuit may look complex, in fact it consists of only three parts: The tone generator (made from IC1), the output amplifier to drive the loudspeaker (Q1),and finally the opto parts. The exact layout of the components is not critical because all of the operating frequencies are low, but it is recommended that you follow the illustration as much as possible to keep the wiring simple. The motor should rotate in a counter-clockwise direction when viewed from above. If it rotates in a clockwise direction, reverse the wires that are connected to the motor.

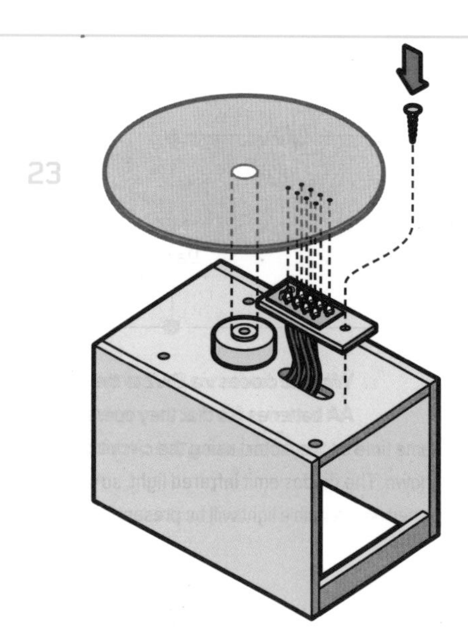

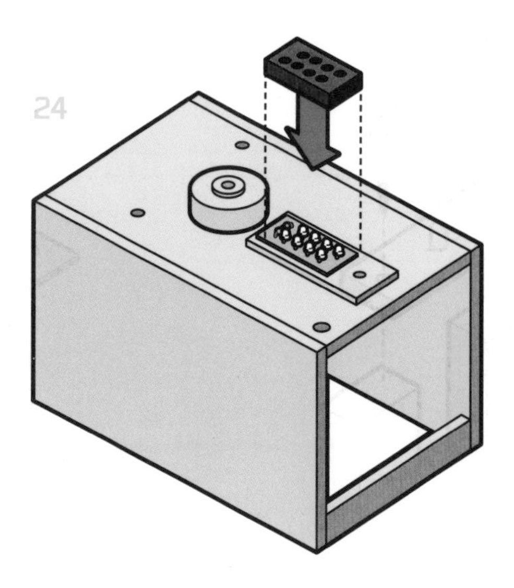

23 Next, align the holes on the test CD with the diodes and use a ½-inch (12-mm) no. 4 wood screw to secure the photodiode array to the top of the box.

24 A shield block is a final addition needed for the diode array (to make certain that only light from the holes in the CD reaches the diodes). Cut piece R to 1¼ x ⅝ inches (30 x 15mm) from ⅛-inch (3-mm) MDF. Paint the MDF black to stop reflections. Drill a pattern of eight ⅛-inch (3.5-mm) holes on piece R, using a piece of matrix board as a jig. The holes should correspond to the position of the diodes. Push this shield block onto the array, as shown.

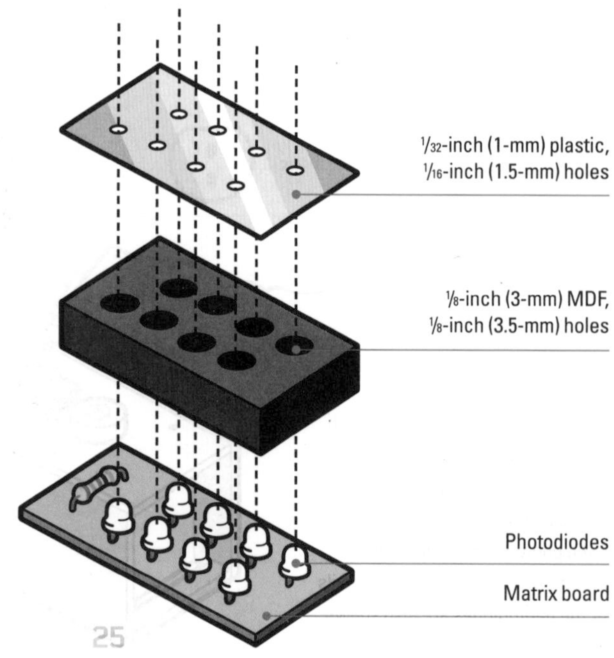

25 Next, drill 1/16-inch (1.5-mm) holes to the same pattern on a small piece of 1/32-inch- (1-mm) thick black plastic sheet, then use impact adhesive to glue this precisely over the photodiodes to stop the light from the illuminator spreading and operating two diodes at the same time. The complete shielding device is shown in exploded form on the right.

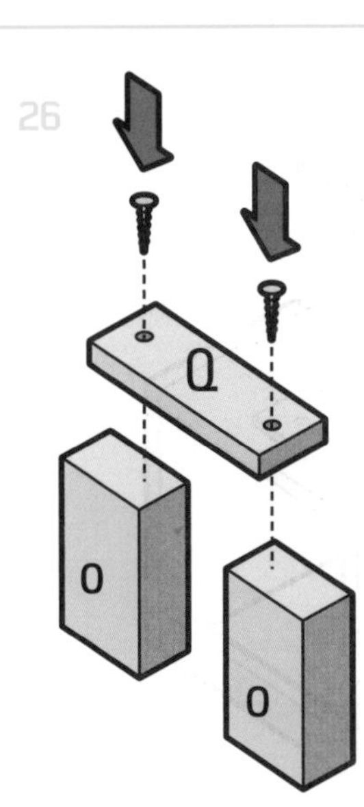

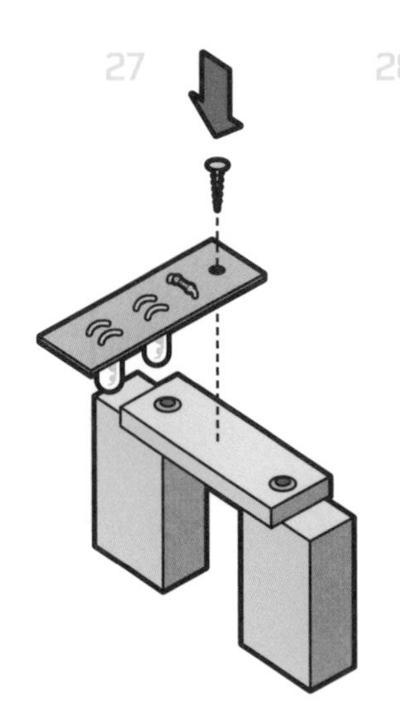

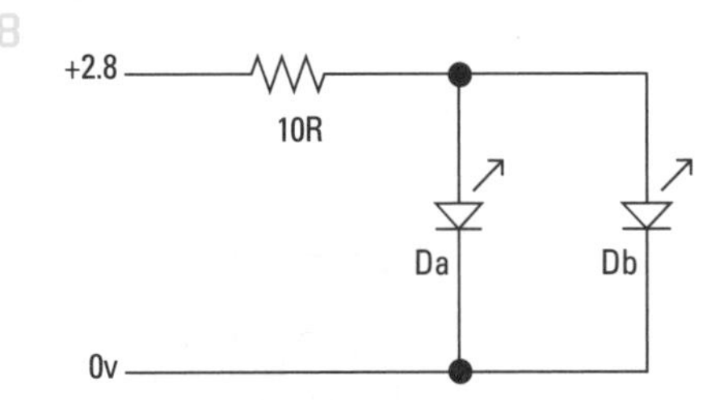

28 Wire the diodes via SW2 to the two AA batteries (so that they operate at the same time as the motor) using the circuit diagram shown. The diodes emit infrared light, so when turned on, no visible light will be present.

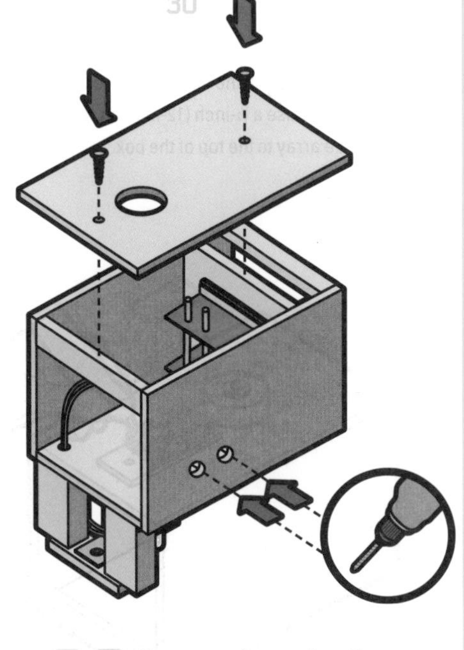

26 You now need to construct a unit from which a light source can be sensed by the photodiodes. Cut pieces O to make two blocks, each 1⅛× 2 inches (30 × 50mm), from ½-inch (12-mm) MDF. Cut 2⅜ × ⅝ inches (60 × 15mm) from 5⁄16-inch (8-mm) MDF to make piece Q and use two ½-inch (12-mm) no. 4 wood screws to screw it onto the two blocks to make an inverted U shape, as shown.

27 Mount two diodes (these function as emitters and evenly illuminate all eight diodes), spaced ½ inches (12mm) apart on a ⅝ × 2-inch (15 × 50-mm) piece of matrix board and use one ½-inch (12-mm) no. 4 wood screw to attach this to the 5⁄16-inch- (8-mm) thick block of MDF.

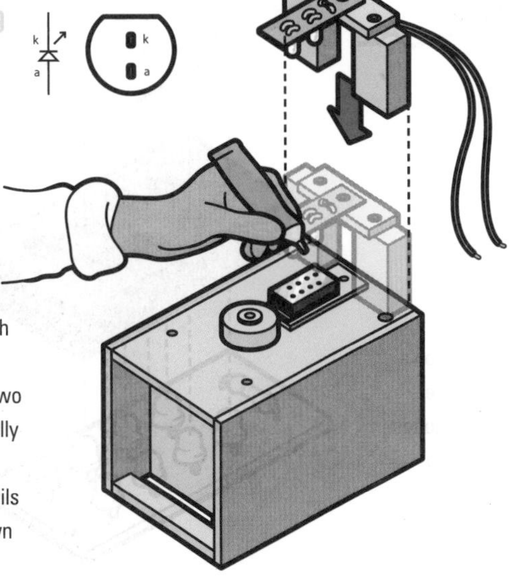

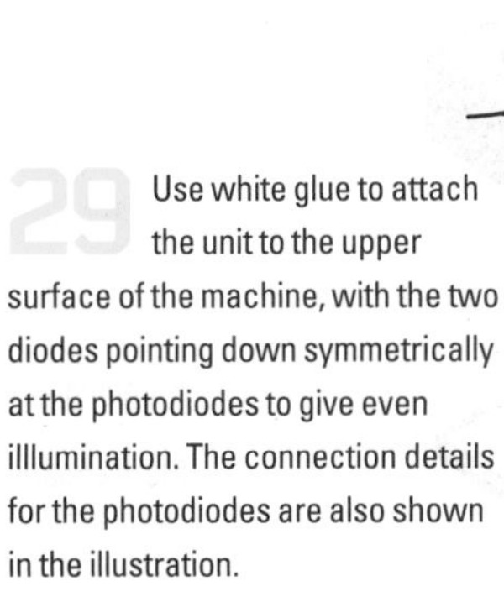

29 Use white glue to attach the unit to the upper surface of the machine, with the two diodes pointing down symmetrically at the photodiodes to give even illumination. The connection details for the photodiodes are also shown in the illustration.

30 Now, screw bottom piece D to inner piece H through two ⅛-inch (4-mm) holes. These holes should be marked out by lining this part up against the bottom of the box; this will get rid of any variations in dimensions that may have occurred during building. Fasten with two ½-inch (12-mm) no. 4 wood screws. Drill two holes on the side of the box for the two switches.

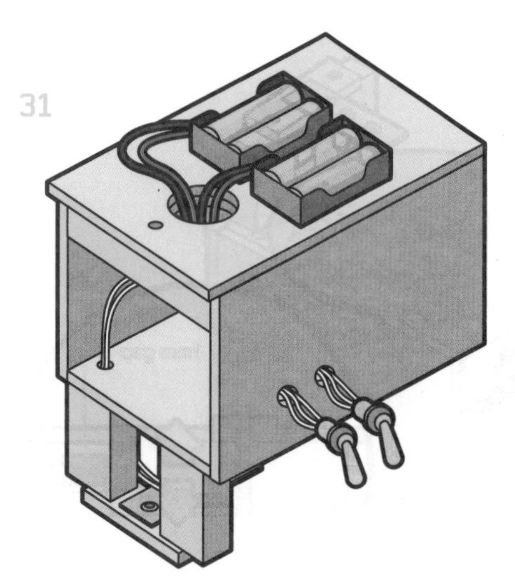

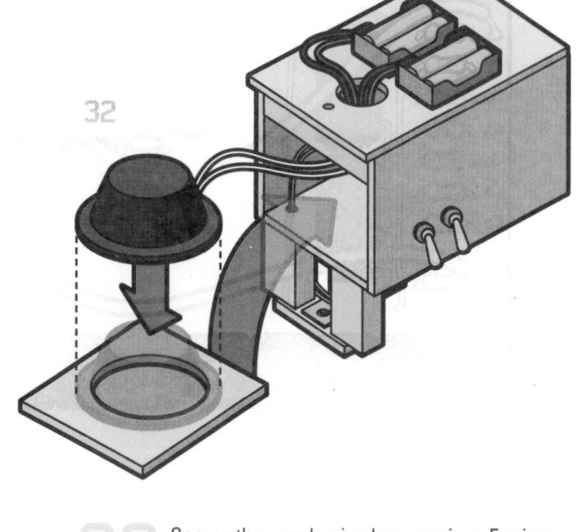

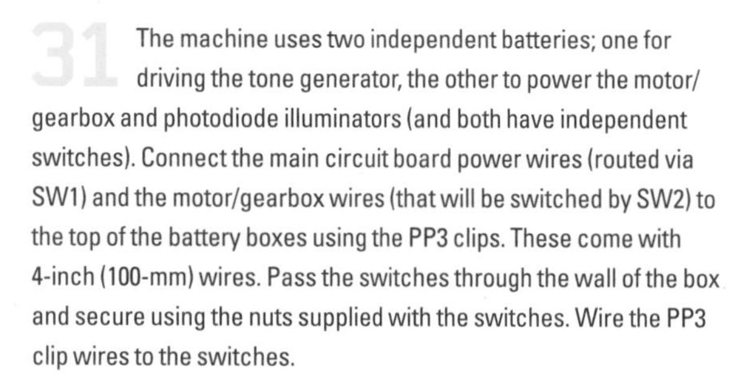
31 The machine uses two independent batteries; one for driving the tone generator, the other to power the motor/gearbox and photodiode illuminators (and both have independent switches). Connect the main circuit board power wires (routed via SW1) and the motor/gearbox wires (that will be switched by SW2) to the top of the battery boxes using the PP3 clips. These come with 4-inch (100-mm) wires. Pass the switches through the wall of the box and secure using the nuts supplied with the switches. Wire the PP3 clip wires to the switches.

32 Secure the speaker in place on piece F using impact adhesive or four screws depending on the speaker's construction. Connect it to the circuit board using two lengths of the PVC-insulated wire. Fasten piece F to the box with two ½-inch (12-mm) no. 4 wood screws.

33 The time has come to breath life into the machine: the tuning stage. Assuming you connected the photodiodes to the circuit board as indicated in step 23, setting the frequency of each note only needs a connection to be made from the positive rail (2.8v) and any of the points marked C D E F G A B. Find middle C (440 Hz) on a musical instrument, tuning device, or frequency meter to set up the first note (the ear is surprisingly sensitive to errors in pitch and the scale is easy to set up once the first note has been established). If you have a frequency meter, then the tuning may be completed in about one minute. The frequencies for each of the notes are shown in the table.

Note	Frequency (Hz)
C	440
D	494
E	554
F	588
G	659
A	740
B	830

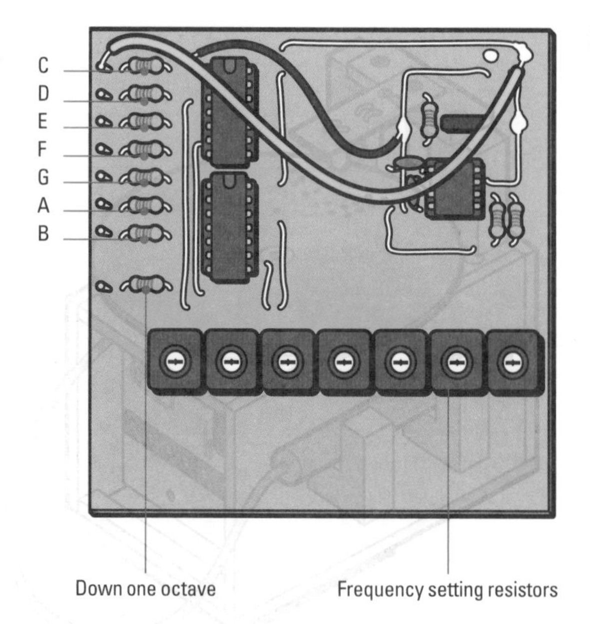

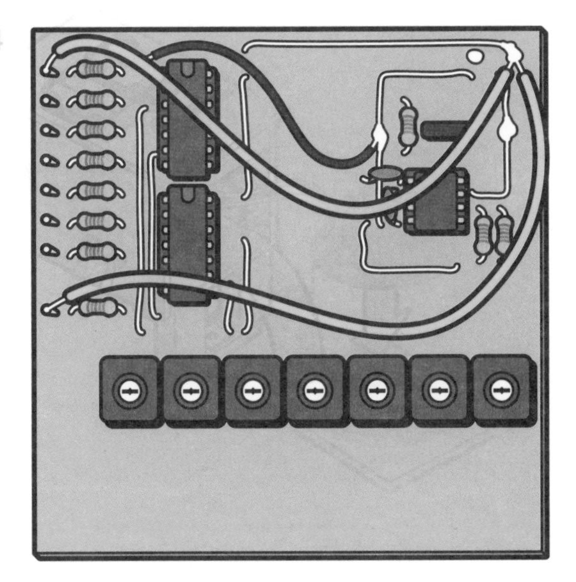

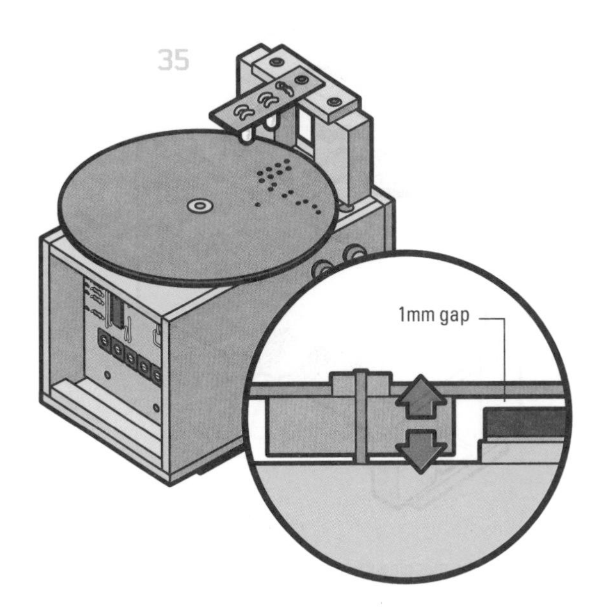

34 Check the operation of the "down one octave" function by joining the C terminal to the positive (2.8v) supply, thus producing middle C, and then connecting the "down one octave" pin to the positive supply as well. The note should halve in frequency, a feature that the ear can easily recognize.

35 Use impact adhesive to attach the drive wheel in place. Before it sets, adjust the drive wheel, moving it up or down the output shaft of the motor/gearbox to provide a critical 1/32-inch (1-mm) gap between the lower surface of the CD and the upper surface of the diode array.

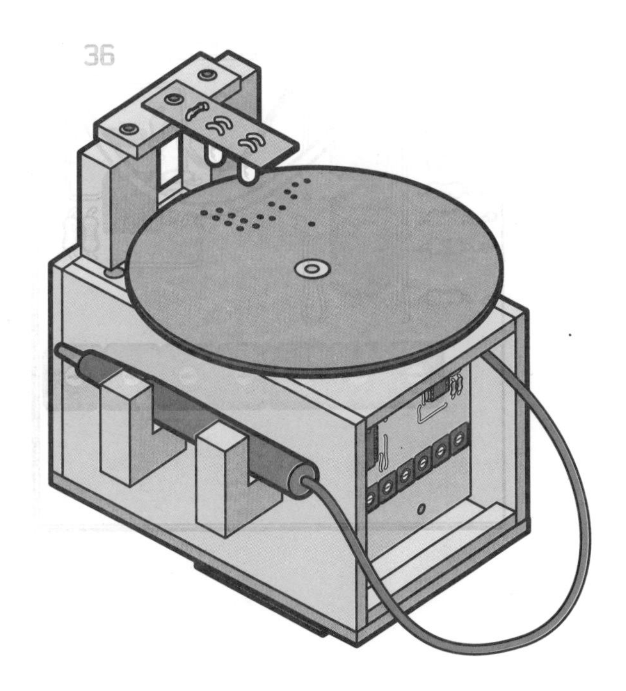

36 Those with a musical ear will be able to write the sequence of notes needed for a tune and program the CD without hearing the notes. Those with a less perfect musical education will want to listen to each note produced by the machine and to write down the sequence. Fit a probe on the machine (in this case from an old voltmeter (although a simple piece of wire will do), and connect this to the +2.8v supply via any convenient point on the circuit board. The probe will produce the selected tone when it touches the points marked C to B. To make a holder for the probe, cut piece P from 1/2-inch (12-mm) MDF, then cut it into two L shapes and secure them to the side of the music box using white glue.

How to Use It

Now the creative part of the project can begin. Your electronic music box is almost finished—all you have to do is decide on a tune. If you write down all of the notes that you want to play, then a simple table, like the one shown right, will provide an excellent way of reminding yourself.

Time slot	C	D	E	F	G	A	B	Octave shift
1	•							
2			•					
3			•					
4		•						
5					•			
6				•				•
57								
58								

Marking Out the Hole Positions

When you are ready to enter your tune, place a blank (undrilled) CD, label side up, into the programming jig. Move the CD, one time slot at a time, and lightly mark the selected notes with a 1/16-inch (1.5-mm) drill bit, using the drilled block to locate the notes.

Drilling Your Hole Positions

Once all your hole positions have been marked out, remove the CD from the jig and drill the holes through the CD using a 1/16-inch (1.5-mm or 2-mm) drill bit. For the "Octave shift" holes, it is always best to use a 1/16-inch (2-mm) drill bit so that the octave is selected before you turn on the tone. When this job is complete, again remove the burrrs from the holes.

Checking Your Program Holes

Place the CD back in the jig and rotate it past each time slot while looking through the eight holes. Wherever a CD has a hole programmed, you should be able to see right through the jig when held up to the light, which is a simple way of checking that all of your program holes have been drilled. (If occasionally you find that you have drilled a hole in the wrong place, no problem; just cover it with a small piece of black PVC tape, and redrill in the correct position). Your electronic music box is now finished and ready for use. Switch on, insert a programmed CD, and enjoy.

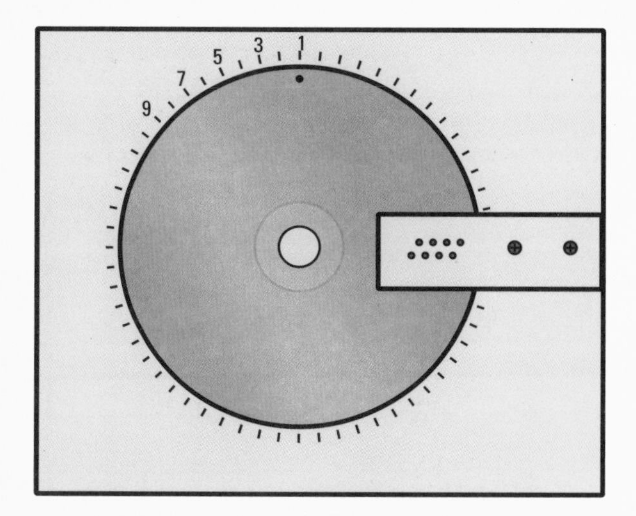

A front and back view of the front of the Electronic Music box.

PROJECT 2

Pyrotechnic Rocket

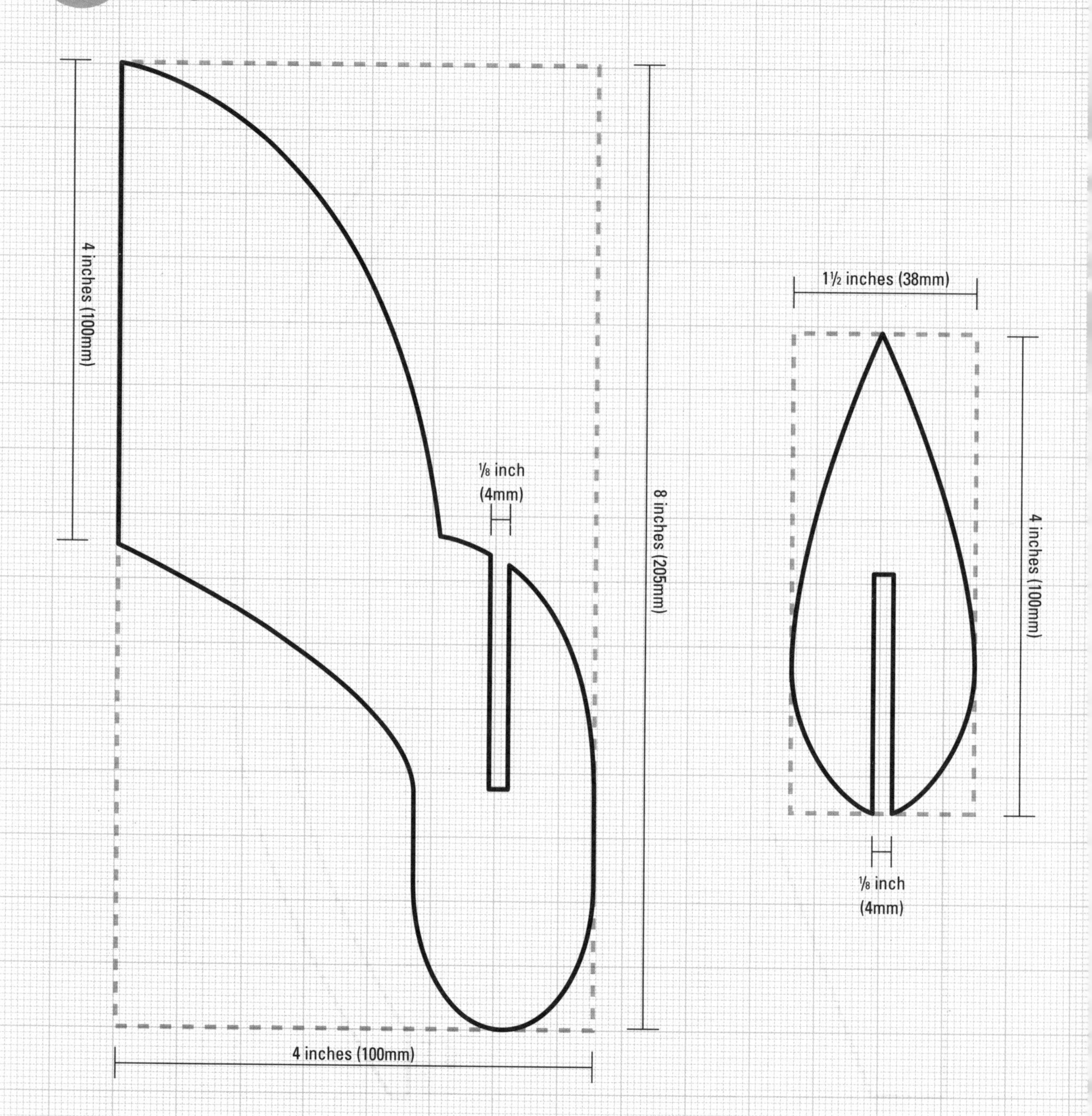

PROJECT 6

Paddle-operated Punt

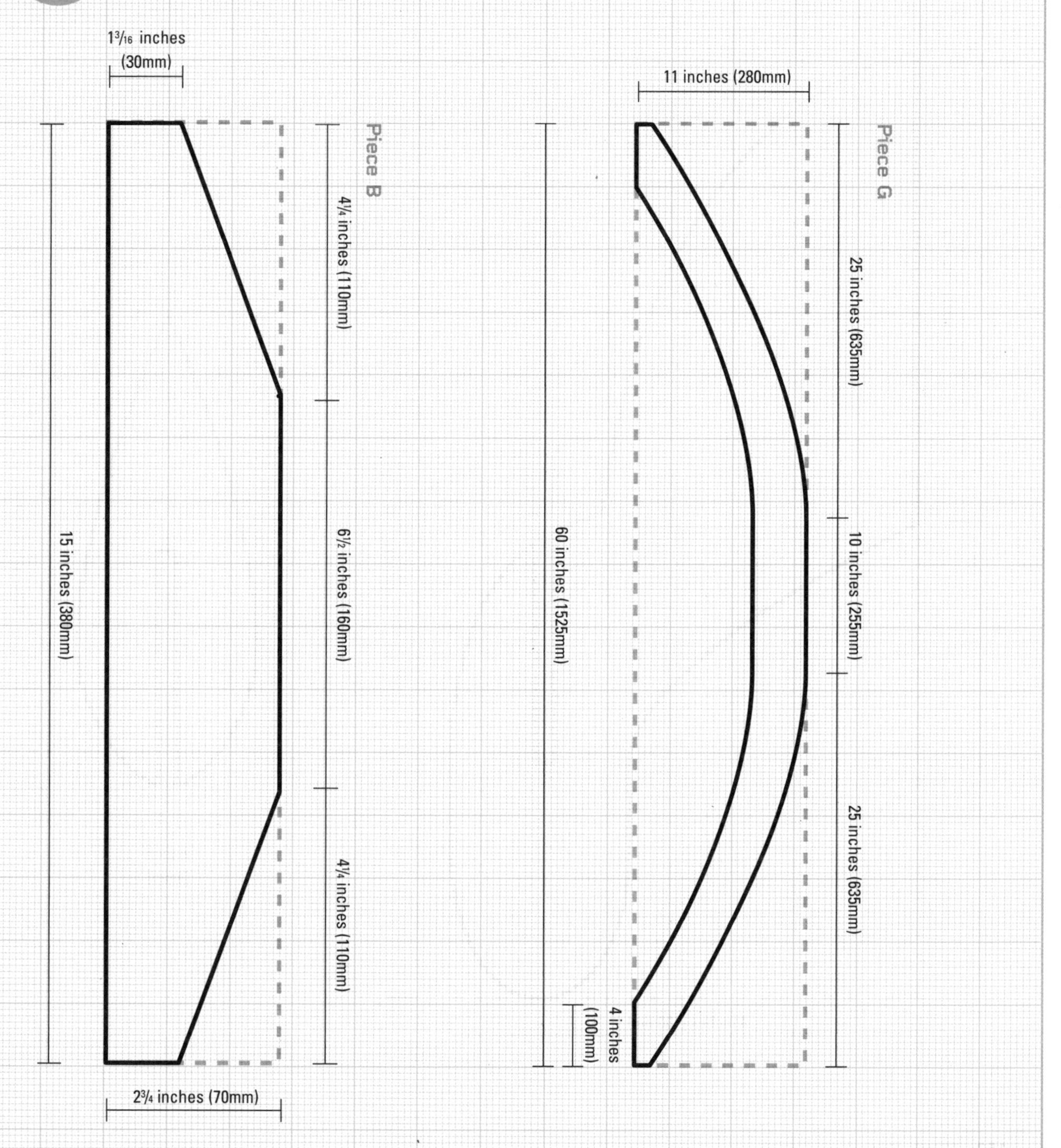

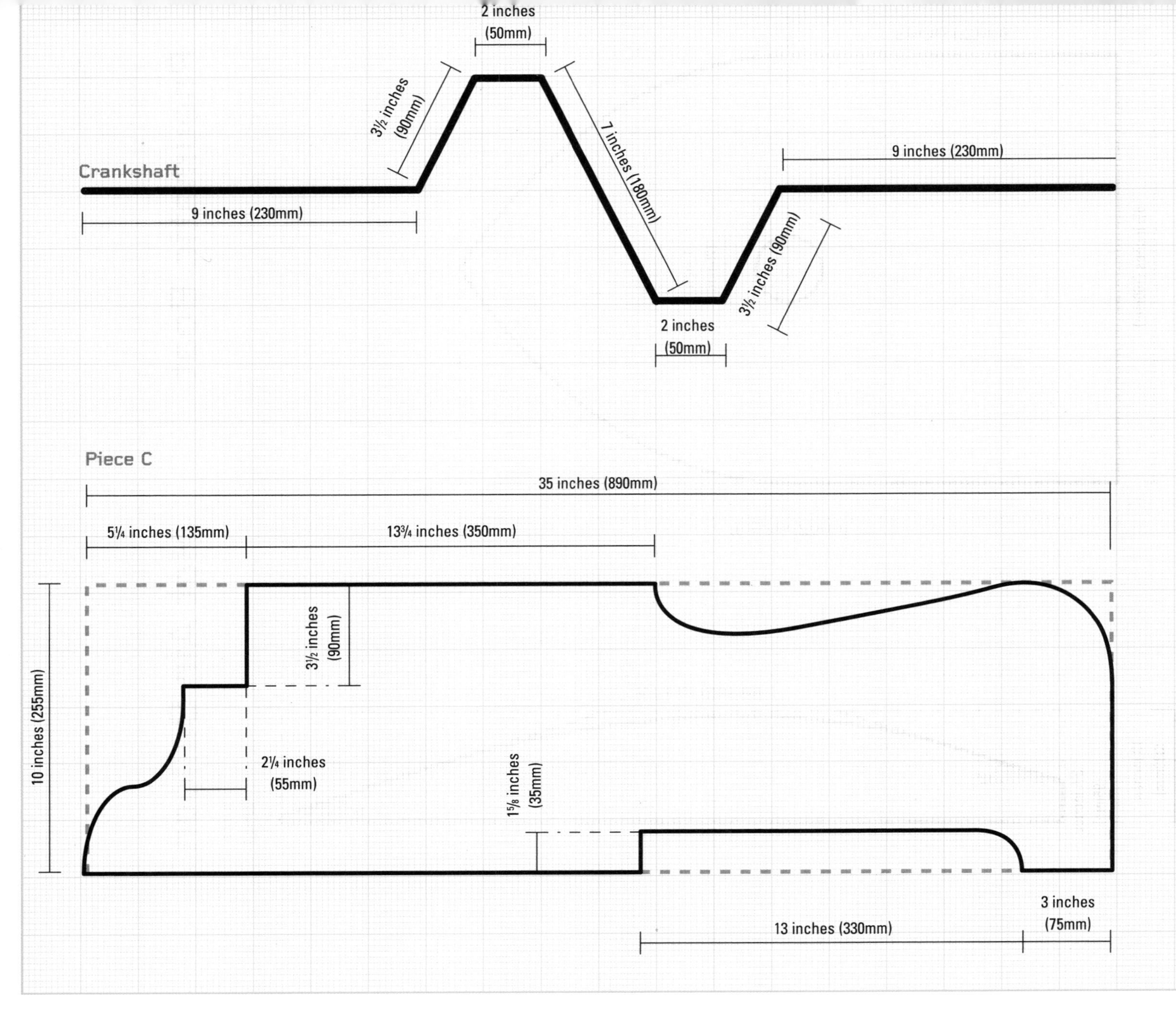

Crankshaft
9 inches (230mm)
3½ inches (90mm)
2 inches (50mm)
7 inches (180mm)
2 inches (50mm)
3½ inches (90mm)
9 inches (230mm)
Piece C
35 inches (890mm)
5¼ inches (135mm)
13¾ inches (350mm)
10 inches (255mm)
3½ inches (90mm)
2¼ inches (55mm)
1⅝ inches (35mm)
13 inches (330mm)
3 inches (75mm)

Pneumatic Boat

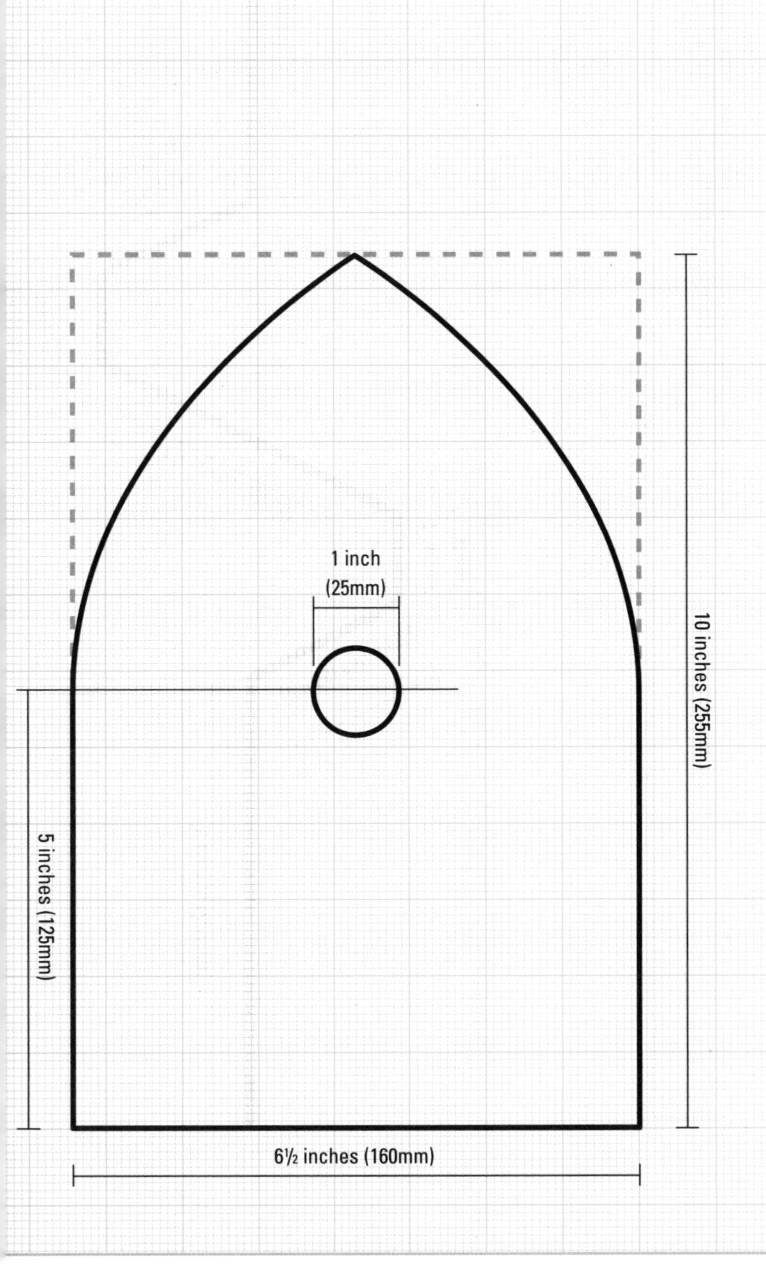

PROJECT 8

Hot Air Balloon

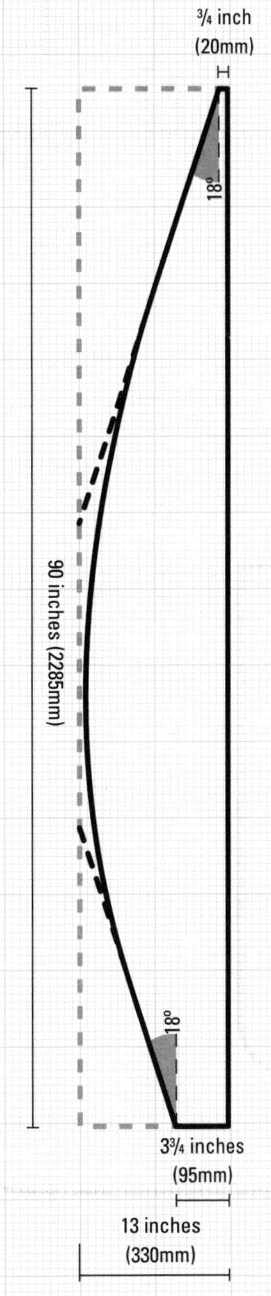

PROJECT 11

Boiler Piston Car

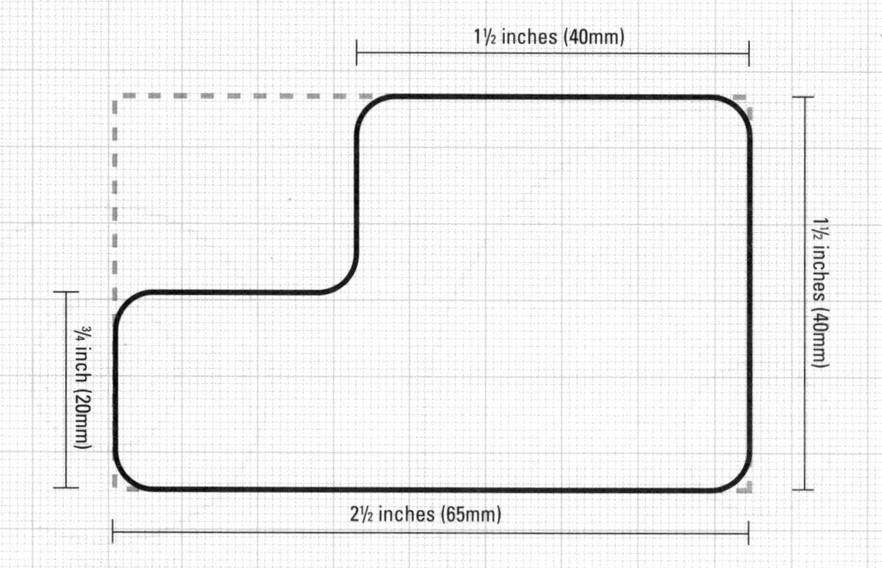

PROJECT 17

Compact Theremin

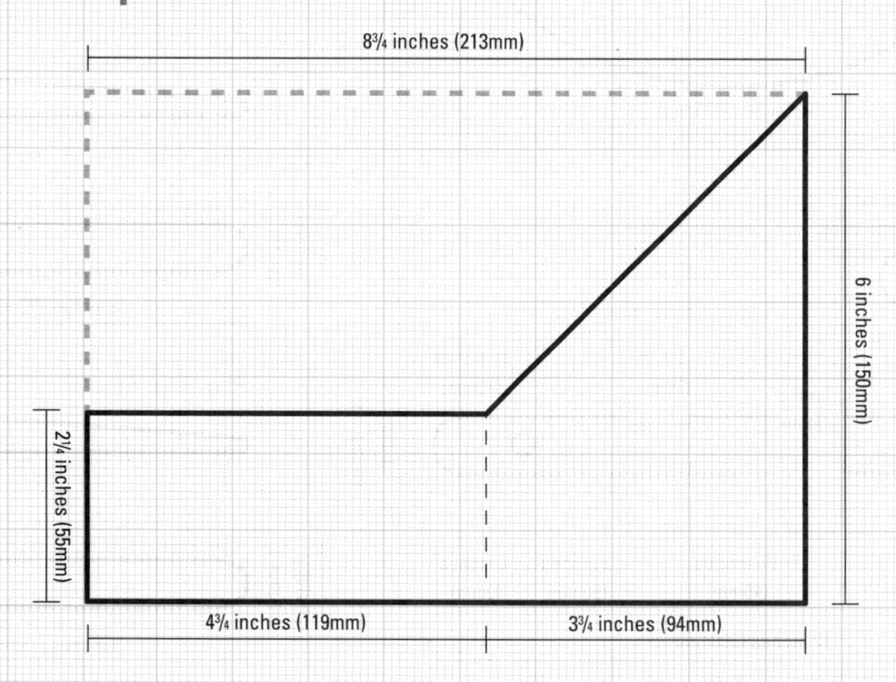

PROJECT 16 Drum Beating Whirligig

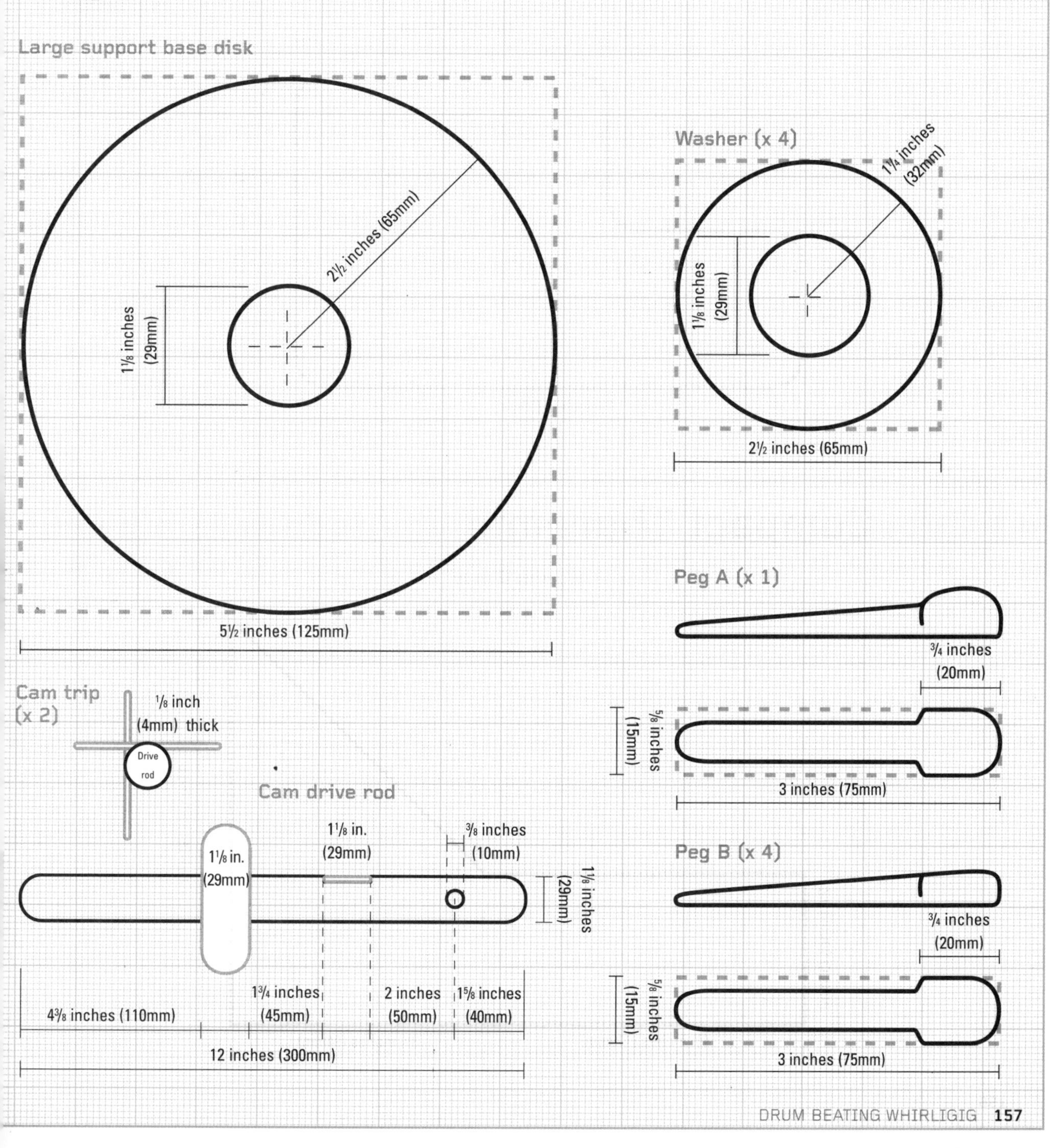

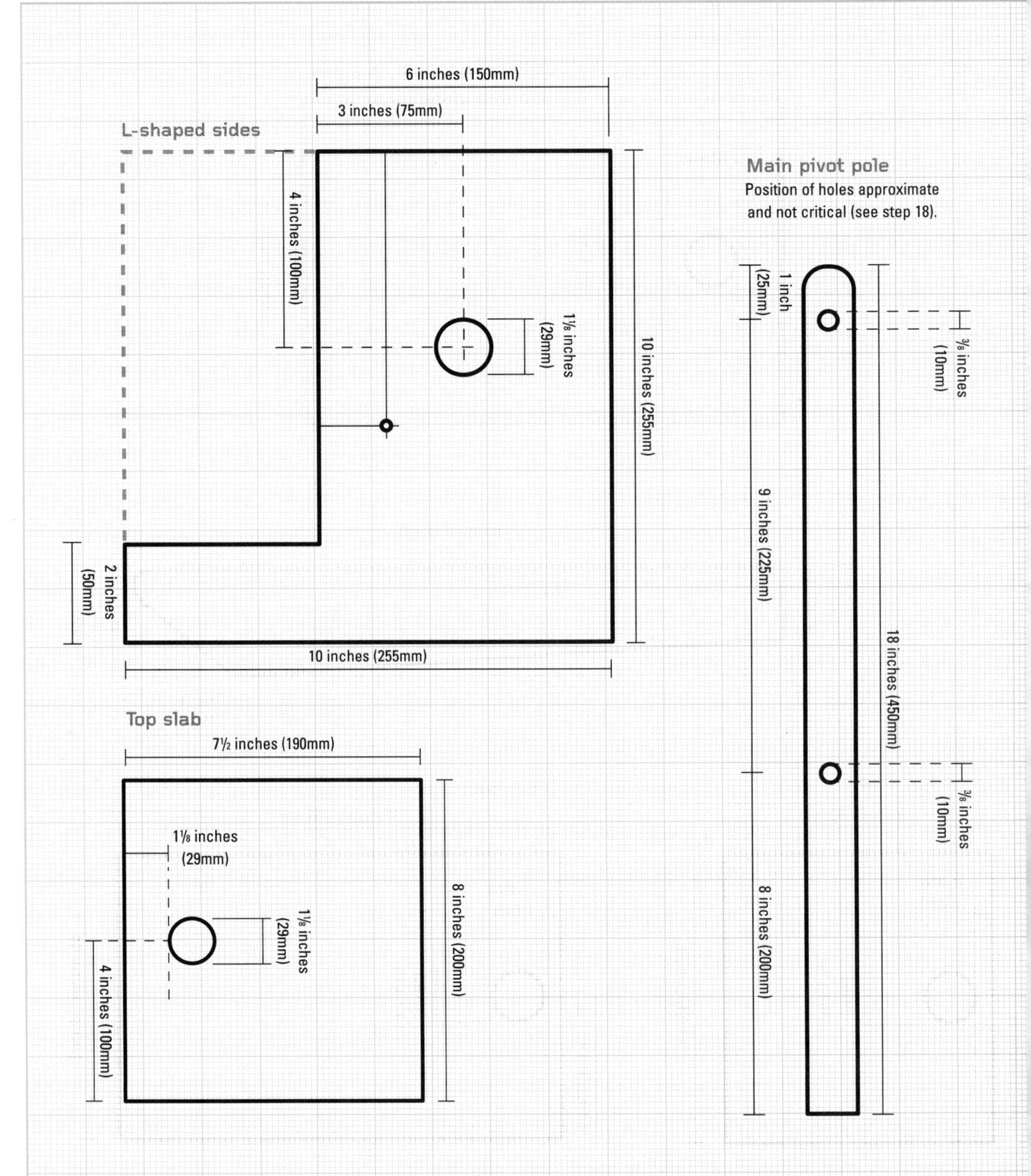
L-shaped sides
6 inches (150mm)
3 inches (75mm)
4 inches (100mm)
1⅛ inches
(29mm)
10 inches (255mm)
2 inches
(50mm)
10 inches (255mm)
Main pivot pole
Position of holes approximate
and not critical (see step 18).
1 inch
(25mm)
⅜ inches
(10mm)
9 inches (225mm)
18 inches (450mm)
⅜ inches
(10mm)
8 inches (200mm)
Top slab
7½ inches (190mm)
1⅛ inches
(29mm)
1⅛ inches
(29mm)
8 inches (200mm)
4 inches (100mm)

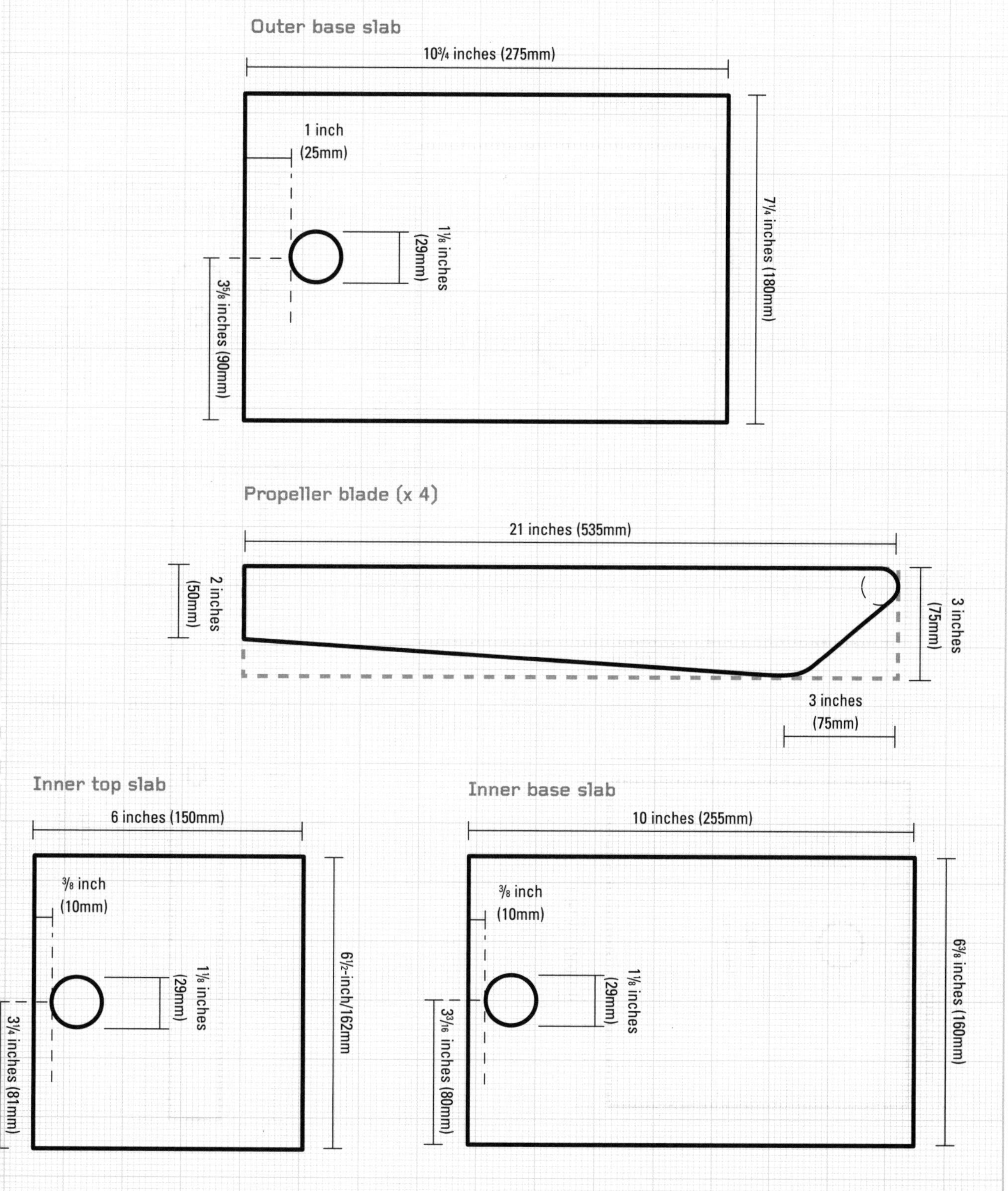
Outer base slab
10¾ inches (275mm)
1 inch
(25mm)
1⅛ inches
(29mm)
7¼ inches (180mm)
3⅝ inches (90mm)
Propeller blade (x 4)
21 inches (535mm)
2 inches
(50mm)
3 inches
(75mm)
3 inches
(75mm)
Inner top slab
6 inches (150mm)
⅜ inch
(10mm)
1⅛ inches
(29mm)
6½-inch/162mm
3¼ inches (81mm)
Inner base slab
10 inches (255mm)
⅜ inch
(10mm)
1⅛ inches
(29mm)
3³⁄₁₆ inches (80mm)
6⅜ inches (160mm)

Index